DISCOVERING DNA

Meditations on Genetics
and a History of the Science

DISCOVERING DNA

Meditations on Genetics and a History of the Science

N. A. Tiley

VAN NOSTRAND REINHOLD COMPANY
NEW YORK CINCINNATI TORONTO LONDON MELBOURNE

Library of Congress Catalog Card Number: 82-4771
ISBN: 0-442-26260-4

Manufactured in the United States of America

Published by Van Nostrand Reinhold Company Inc.
135 West 50th Street, New York, N.Y. 10020

Van Nostrand Reinhold Publishing
1410 Birchmount Road
Scarborough, Ontario M1P 2E7, Canada

Van Nostrand Reinhold
480 Latrobe Street
Melbourne, Victoria 3000, Australia

Van Nostrand Reinhold Company Limited
Molly Millars Lane
Wokingham, Berkshire, England

15 14 13 12 11 10 9 8 7 6 5 4 3 2 1

Library of Congress Cataloging in Publication Data
Tiley, Nancy.
Discovering DNA.

Bibliography: p.
Includes index.
1. Genetics. 2. Genetics—History. I. Title.
II. Title: Discovering D.N.A. [DNLM: 1. Genetics—History. 2. Genetics. QH 430 T572d]
QH430.T54 575.1 82-4771
ISBN 0-442-26260-4 AACR2

Preface

The subject of genetics first began to absorb my attention in the early 1970s. The fervor of intrigue and ethical indignation which I then felt have both remained and changed, becoming balanced with appreciation, through the many forms of early expression which culminated in the present work.

For their help in my endeavor as one of the many who have sought to bring the subject of genetics and its history to a clearer understanding for those persons not directly involved in the science, and as one of those who have perceived ancient themes in the subject, I wish to thank, among others, Arthur Caplan and the Institute of Society, Ethics and the Life Sciences (Hastings Center) for opening the doors of the institute to me and providing large quantities of enormously helpful material regarding ethical issues in genetics. I am indebted to the library of the New York Academy of Medicine and to the New York Public Library. For their conceptual, editorial, or other assistance in the preparation of the manuscript I especially thank my family, and my friends, R. de Marrais, S. Roy, E. H. Russell, J. C., J. Michell, F. Hogman, L. W., L. G.; and J. Burke. My thanks to N. C., especially, and to S. G.

Introduction

"The greatest single achievement of nature to date was surely the invention of DNA. We have had it from the very beginning, built into the first cell to emerge, membranes and all, somewhere in the soupy water of the cooling planet."—Lewis Thomas

Whether our origin lies in the manifestation of Idea, in the actualization of sacred decree, or in a version of Darwinian evolution, the gene presides over our physical character. The stuff of life? DNA!

Biology has acquired the ability to alter existing life forms and to create new ones through genetic engineering. When we speak of creation and evolution, we must also speak of such laboratory creations as the "oil-eater," a bacterial "bug" never before seen on the face of the earth. New life formation is just one of the subjects explored by a whole new breed of scientists. Medically oriented research is an enormous area of promise. Work in the research of diseases provides astounding results, conclusions, and hypotheses. Human insulin is synthesized; new and safer vaccines are created; hormones are mass produced by bacteria.

Coincidental to studies of DNA and its manipulation are studies in form and the symbolism of form, especially with consideration of the double helix of DNA. Modern research does not have isolated effects. It opens new vistas in understanding and dealing with medical and technological problems. It also opens to views on ancient and continuous problems such as those revealed in parallels between social and biological forms of communication.

The discovery of details about the "stuff of life," how it functions and what can be done with it has enjoyed rapid progress since the rediscovery of Mendel's laws of heredity: From 1900, genetics became a science in its own right. Its development required a multitude of contributions from various fields, including mathematics and physics. The work led, by the 1970s, to a controversy in which scientists themselves began to doubt the safety and ethics of their work. To possess DNA is to possess life, and it might be said that to be able to alter DNA is to hold power over life. Despite the controversy, the work has continued. Indeed, the controversy served largely to bring studies in genetics into greater recognition. Genetics is now recognized as one of the most rapidly developing areas of scientific research.

How was DNA discovered; by whom? What dreams still embrace its discovery? What realities prove to be technological/theoretical limitations falling by the wayside as the legitimacy of almost the entire "dream" spectrum is realized? Finally, what is the DNA controversy all about?

Many years ago a young German discovered something he called "nuclein." The "nuclein" was DNA but the young man did not know this and thought the substance inconsequential. Centuries had been required to recognize that the cell is responsible for heredity, that the gene within the cell is the actual carrier of hereditary information. But it took millenia simply to conclude that heredity exists at all, that the formation of the child emerging from a woman has depended on a man's involvement with the woman; that the child does not simply emerge, it is born, and it is born in the likeness of its parents by no simple coincidence of the fates.

"Microbiology, like all the sciences, is founded upon the twin pillars of craft technique and philosophical speculation."—Patrick Collard

Several minor notes might be made in regard to the preparation of the manuscript. Biological communication is important in terms of genetics but also in terms of the functions of the brain. The brain is of special interest as we enter the age of the computer that mimicks it and in terms of the behavior modification potentially affecting it through techniques derived from studies in genetics, therefore the fundamental cellular basis of communication for the brain is explored rather more than it may otherwise have been. In regard to the word *evolution,* which is used not infrequently, there is attached the general definition of "growth," or simply "change," unless the subject of Darwinian evolution is being directly discussed.

"To put the work of Watson and Crick in perspective and, indeed, to appreciate fully its importance, it is necessary to consider the long history of ideas and of experimental work on the chemical nature and three-dimensional structure of both types of naturally occurring nucleic acids and also on the nature and basis of heredity in living organisms."—Lord Todd in his Foreword to *A Century of DNA*

L. C. Dunn has written a splendid little volume entitled *A Short History of Genetics,* of tremendous value to historians. Dunn, however, exhibits a limited time perspective on the subject, and his argument against a more extended perspective must be dealt with. He has written that "to trace the origins of genetics to a time before the beginnings of modern science seems to me to lose sight of its essence. To begin with Aristotle is like beginning a history of some aspect of literature with the origins of the alphabet or of language. Here, I take a narrow view and do not consider the science of genetics equivalent to thinking about its subject matter—heredity and variation." However, I have gone as far back into antiquity as the Cro-Magnon, and have found the various foundations for the "literature" of great interest for a clear understanding and balanced perspective on the modern situation.

The studies performed by a variety of historians, including the first historians, the mythologists, have assisted this work. Mythology, with its metaphors, is particularly appropriate to genetics when we consider Gregory Bateson's view of the role of metaphor in biology. Bateson, in "Men Are Grass," the last lecture* he delivered to the Lindisfarne Fellows before his death, describes two forms of syllogism. One of these is the *Barbara,* or Socrates syllogism (men die; Socrates is a man; Socrates will die), which, requiring subjects, "could never be much use in a biological world until the invention of language and the separation of subjects from predicates." The other, Bateson's kind of syllogism, is less limited and more metaphorical. It is the grass syllogism (grass

* A transcript of the lecture is given in the *Lindisfarne Letter,* no. 11, June 9, 1980.

dies; men die; men are grass), which is concerned with affirming the consequent. Bateson suggests that "it looks as though until one thousand years ago, perhaps at most one million years ago . . . there were shared predicates between the horse and the man, which zoologists today call homology." He goes on in his defense of the metaphor and the relevant logic of the grass syllogism for students of biology. Bateson concludes that "metaphor . . . was in fact the logic upon which the biological world had been built, the main characteristic and organizing glue of this world of mental processes. . . . " In genetic engineering we find a subject both highly praised and highly controversial, a subject providing, additionally, literal and metaphorical insights.

Through studies of the gene tremendous insights into biological operations have been gained and remarkable, even fantastic, manipulative results have been achieved. Controversially, genetic engineering has witnessed both practical and philosophical dilemmas which will require years to be completely resolved. Very few writers have alluded to the philosophical dimensions as beneficial or, at the least, instructive. While there are very real and reasonable philosophical issues complicating acceptance of genetic engineering for both those of the inner circle and those outside it, there also is the possibility of assistance to philosophy and to society in general in ways doubtless unrecognized by most genetic engineers. One of these benefits is found in the symbolic spiral structure reflected in the double helix of DNA.

The double helical structure of the DNA molecule is, at least in part, nonpolar, an attribute common among biological molecules. Yet the nature of alignment between the strands of DNA defines, to a large extent, the nature of life for an organism. If they result in a confusion or distortion of characteristics the strands become analogous to stresses in the relationship of two poles. These two poles are described in abstract terms by geometrical symbolism concerned with the quality of nonmaterial life; this symbolism may apply to material life as well. The double helix has a structural symbolism affording comparative similarity with many staff-of-life symbols. This is especially true with one of the symbols, the caduceus, sign of the physician holding power over life and death with the staff of harmonized polarities. Studies in genetics are coincidental studies of ancient symbols, of metaphorical bridges between many different studies.

". . . The principle of life consists in the tension which connects spirit with the realm of matter, symbiotically joined. The element of life is dominant in the very texture of nature, thus rendering life, biology, a transempirical science."—Ruth Nanda Anshen

Now that society has strengthened the various facets of a grand differentiation it has begun to strive toward their reintegration. I have been conscious of those beginnings in the preparation of this volume. Studies in genetics stem from a desire to heal. The results of research in genetics are geared primarily to the medical world. Our focus, then, begins in the world of the healer.

Medicine and Medical Myths

"The reason is of ancient date why myths have come, in vulgar estimation, to be synonymous with lies; though true myths—and there are many—are the most comprehensive and splendid statement of truth known to man."—Jeremiah Curtin, Myths and Folklore of Ireland

As a note to the following myths regarding healers and medicinals it is necessary to be reminded that two definitions of the word "myth" exist. Each of these is, in turn, implied here. In the first instance, with medicine myths, the word is used in reference to beliefs of a nature superseding the perceptual orientation of the physical and chemical sciences. Yet the analogies between the physical and symbolic worlds are necessarily extensive. Such analogies are evident in the statement that "if Levi-Strauss is right, myths are constructed by a universal logic, like language itself, [and mythologizing] is as characteristic for human beings as nest-building is for birds" (Lewis Thomas).

The more recent definition of "myth" is employed in the section on medical myths, and is associated with fallacy. Comprehending the difference between fallacy and symbolism is essential to understanding the difference between the two definitions for the word "myth".

We will begin with a traditional orientation, symbolic mythology. In mythology we find not only practice in percieving metaphor but also intriguing perspectives on the healer and the art of healing.

MEDICINE MYTHS

The following is a combined view of versions of the origin myth of the Medicine Rite of the Aboriginal Winnebago as related by Paul Radin, Special Publications of Bolligen Foundation, No. 2, 1950.*

"Our father, on what he must have been sitting (when) he came to consciousness? Then his tears flowed (and) he wept. Not long did he think, not as anything did he see, not anything was there anywhere. On what he thought he must have been sitting, from that seat, he took something. From that something which he took (from a little piece of it), he made this world. . . . He [another he] went out of the lodge in which they lived at night and wandered about. Within that little lodge, whenever he walked about, he spread light within it. When daylight came, he became quiet, it is said. When the sun went down he went out (again) and travelled over the full extent of our grandmother [earth], all over her he wandered. Then, early, before sunrise, he came in (the lodge) and was quiet. Throughout the day he kept thinking of the work he had to do. . . . 'Grandmother, it is good and you are going to be of help. Grandmother, you may now close up your breasts.' . . . Thereupon Hare got up again, it is said, and spoke, it is said, thus: 'My friends, for my uncles and my aunts, it is just this that I have been desiring.' . . . There, the whole long day, they taught the people how (to obtain life) and only when the sun had reached to the top of the trees, when the time for terminating (had come), did the spirits disperse. When they left for their homes they took with them half of the light-and-life. They rubbed against the door-entrance posts as they went out and pushed these deep down (into the earth) so that they would never be able to collapse."

In the same world, "the first one was the very white walker-on-light, the very awe-inspiring cackling-swan who had come from above and whom like the other walkers-on-light Earthmaker had placed in control of most abundant life-and-light. He it was who set out to walk on light (to invite the spirits). He made the circuit of the lodge four times and (each time) life-and-light became brighter; the toes on his feet were

* From *Winnebago Culture As Described by Themselves-The Origin Myth of the Medicine Rite: Three Versions-The Historical Origins of the Medicine Rite,* by Paul Radin, Special Publications of Bollingen Foundation, No. 2. 1950: Waverly Press, Inc., Baltimore. Also issued as Memoir 3 of the International Journal of American Linguistics, formally of Indiana University Publications in Anthropology and Linguistics. University of Chicago Press.

not uneven and he stained-and-smeared the fireplace with life-and-light. . . . When he [another he]-whom-they-call-nephew planned to gather together the spirits (for the Medicine Rite) he used a tanned buckskin, and tobacco, the means for obtaining life, that is what he used to gather the spirits together. He-whom-they-call-nephew divided the tobacco into four parts. Then he tied these parts up. He tied life-and-light into them. He was going to ask those (about to start off) to obtain blessings. Each one in turn started out to gather the spirits together. . . . To the great island-anchorer who is in the east Earth-maker gave a messenger (medicine pouch) thus placing life-within-the-lodge (gourd). This messenger was blackened with age and he picked it up and rattled it gently. . . . There at the doorway had been placed (when the lodge was erected) a female blue snake on the right (left?) and a male snake, on the right. He seized ahold of the female snake placed on the left (and the male snake). Both of them he shook violently. For the whole extent of our grandmother and right down to her deepest point, thus far did he push these snakes (the posts of the doorway). There they would remain never to be loosened. Then he pushed the door-flap back and, with it, went half of the light-and-life. Then he made that which stands in the center of the lodge round and he took away (the second) half of the light-and-life. The walkers-on-light made their way out carrying something in their mouths. Those of the earth made their way out carrying something in their mouths."

The "something in their mouths" refers to a concept still in evidence among various medicine men, including those of North American Indians; it implies that the healer has taken the disease demon into his mouth, where it rests, if the healer is practiced and successful, until it is spit out in a way rendering it harmless, at least to the patient it had inhabited. The parentheses used in the foregoing are those of Radin, while the brackets are those of this author.

The myth seems, in a sense, to be foreign to us yet it is little more foreign than the myths provided by North American Indians. I am told that the most desirable of North American Indian medicine myths are not very accessible, particularly at this time. They are considered by the Indians to be of a sacred value requiring containment from possible dilution or contamination or other confused response. This may be highly unfortunate or it may not be. The old mystery schools were very often designed precisely to introduce a greater number of people to their understanding. "In fact," states David Spangler, "the whole function of a mystery school was to demystify the mysterious."* Society may still not be adequately prepared for too intensive an introduction into the views of the mystery schools. Meanwhile, however, preparation toward a future in which such an introduction may indeed be possible must necessarily include a certain measure of familiarity with the subject. The subject is one of interest on various levels, including the history of medicine.

Shamanism is the original profession of the physician. The origin of the shaman (or medicine man) is the ultimate origin of the physician. This origin is viewed throughout mythology. In his *Introduction to a Science of Mythology* Claude Levi-Strauss relates a Warao myth concerning the origin of tobacco and of the first medicine man. The Warao myth** is as follows.

"An Indian called Komatari wanted some tobacco which, at that time, grew on an island out at sea. He first of all approached a man who lived alone on the shore and whom he mistakenly took to be the master of tobacco. A humming-bird joined in their conversation and suggested going to fetch some tobacco leaves. But he made a mistake and brought back tobacco flowers. So the man who lived on the shore set off for the island and succeeded in eluding the people who were keeping watch over the tobacco. He came back with his corial full of leaves and seeds, with which Komatari filled his basket. The stranger took leave of Komatari, but refused to give his name. He said the other would find it out for himself when he became a medicine-man.

"Komatari refused to share the tobacco with his companions. He hung the leaves under the roof of his house and left them in charge of the wasps. The latter allowed themselves to be bribed by a visitor who offered them fish and who stole some of the tobacco leaves. Komatari noticed that some tobacco leaves were missing and drove away all the wasps except one particular kind, which he made his watch-men. Then he cleared a small area in the forest and planted his tobacco.

*Spangler, David, "The Reappearance of Mystery Schools," Lindisfarne Letter 7, Summer 1978.

**M328. "Warao 'The origin of tobacco and of the first medicine-man' (2)" in *From Honey to Ashes* by Claude Lévi-Strauss, translated by John and Doreen Weightman. English translation copyright © 1973 by Harper & Row, Publishers, Inc. and Johnathan Cape Limited. Reprinted by permission of the publisher.

"From four spirits, whom he met in turn but who all refused to tell their names, he obtained the calabash tree, feathers and cotton-twine which were to decorate the first rattle, and the stones which would make it reverberate. Warned by the hero that the finished rattle would be used to destroy them, [these particular] Spirits took their revenge by causing diseases. But to no avail; thanks to the rattle, Komatari cured all the sick people, except one man whose condition was too serious. It would ever be thus: some patients would be saved, others could not be saved. Of course, Komatari now knew the names of all the Spirits. The first one he met who procured the tobacco seeds for him was called auuno (Anura in Arawak) 'White Crane.'"

MEDICAL MYTHS

Mythology is the poetic expression of reality. It is a form of expression we have long forgotten and one which is not readily associated with the root word of its name. The fallacies, on the other hand, within a shifting perspective on the world present a subject very different from the poetry of the mythology described in the previous pages. Now we encounter the word "myth" in its common reference to fallacy. Medicine preparing to enter the twenty-first century encounters the yet prevalent belief that no known cure for the common cold exists, although a major cause of the common cold is known to be the JH virus against which the first successful vaccine was developed by Winston Harvey Price (1923–1981), Professor of Epidemiology, Johns Hopkins University.

"Myth": Stomach problems are a sign of genius.
"Myth": Eating potatoes will lead to spina bifida in infants.
"Myth": Normal human cells growing in a laboratory culture dish were reported in the 1960s and early 1970s to have the exhibited capacity for spontaneous transformation into a cancerous nature (see HeLa cells). (Belief in spontaneous transformation of cells is no longer acceptable to the majority of scientists.)

Medicine, like any other social development, whether that development be a science, an art, or both, has a history of modulated appropriateness. Its effects are not always direct nor are they always most evident within their own framework. When its applications or attitudes are not directly effective they may be found, sometimes humorously, to have important extenuating effects. An example is that "while Dr. Benjamin Rush and the others who insisted on the local origins of disease [the Philadelphia yellow fever epidemic of 1793, etc., was ascribed to local filth] were partially wrong, their mistaken theories nevertheless had value in promoting urban improvements."*

Folk medicine is a trial-and-error search for remedies. Occasionally this empiricism has been fruitful. Withering's discovery of the drug digitalis provides an example of successful empiricism. Folk medicine addresses itself not only to remedies but also to preventative treatments. Another case of a borrowing from folk medicine is found in the introduction of inoculation against smallpox, the first large-scale use of which was directed by Cotton Mather and Dr. Boylston in Boston during the epidemic of 1721. Cinchona bark, used as a folk drug, was one of the first truly specific drugs discovered, and it influenced medical theory and practice throughout Europe.

Sufferers have long had their own remedy, if not cure, for the common cold. The comforts of homemade chicken soup are associated with the realm of home remedies for a variety of afflictions; other similar remedies include onions and the licking of one's wounds. The effectiveness of these remedies exceeds that of the placebo. The soup benefits mucous membranes; onions, used in every manner from the raw state for eating or house fumigation, to the heated state for poultices applied to the chest to loosen congestion, contain a natural form of the antibiotic penicillin; saliva, apart from the fact that it may possibly provide a chemical clue to the development of social behavior, and in spite of the fact that the human mouth contains more germs than the mouths of most animals, does indeed help cuts to heal. The effectiveness of folk remedies is not entirely without legitimate acclaim.

The notoriety of marijuana is not dependent on its ability to relieve nausea and vomiting, but the THC it contains does in fact very nicely accomplish this relief as well as a variety of others including the reduction of eye pressure such as that common in glaucoma. In keeping with

* *Medicine in America,* Shryock, Richard H., editor, 1966, Baltimore: Johns Hopkins University Press.

a form of practicality demanding the expansion of technology, on the fringes of euphoria-denying Victorianism, and due to the illegality of the marijuana containing THC, a new drug was developed. The new drug, Nabilone, successfully mimics noneuphoric THC, that is, it achieves what THC achieves but without the side effects accompanying THC in its natural environment in marijuana. While the medical triumph is obvious, the question Buddhists might justifiably ask is: Is Buddha, often depicted with soma leaves identified with hemp, at all synthesized or quite literally separated from an aspect of his essential portrait for those recipients of the new drug who do not perhaps realize how close they came to touching a Path?

Opium, morphine, heroin, and cocaine were historically distributed by a wide range of breeds of physician who obtained a legal monopoly over them by virtue of the Harrison Narcotics Act of 1914, before which time the opiates were legal, inexpensive, and available to all daring their responsibility. Continuing a precarious course toward establishing the medical profession preparation of opiates after 1914 was as seemingly indiscriminate as their availability; this was the case in preparations of nonopiatic medicine as well. The potentially debilitating side effects of many contemporary medicines and medical treatment is sadly reminiscent of the effects of many earlier prescriptions. In at least one instance a medicine has increased the specific problem it was intended to remedy. The Holloways' Pills of the Civil War were sold to victims of dysentery who purchased, with later regret, a powerful purgative which only accomplished what dysentery was already doing far too well. This is an extreme example of poorly prescribed treatment.

The breed of physicians out of which as much productivity as superfluous activity is likely to rise is not absent at the curve of a turn into a new century. DNA and RNA pills and diets to vitalize, extend, and further qualify one's life are a common and recent addition to vitamin displays and druggists' counters. That diets of "DNA-rich" salamanders (coincidentally important in metaphysics) are as unpleasant as a dish full of worms, which form the real if rare constituents of a few modern diets, or, that DNA/RNA pills accomplish little benefit save the moistening of the internal regions as they harmlessly dissolve and are passed, is of no practical value to the hopeful willing to try anything. Powwowing would likely be more effective than such diets.

Powwowing, the combined practices of the herbalist, spiritualist, and intuitive magician, was an art most familiar to the first generations of America's Founding Fathers and one of lesser prevalence in the memories of the elderly of the late twentieth century. Both realistic and superstitious, and altogether innocent in terms of what was known in chemistry and medicine a hundred years ago, a slim 1819 volume entitled *Pow-Wows or, the Long-Lost Friend* (John George Hohman) includes A Good Way To Cause Children To Cut Their Teeth Without Pain (application of boiled rabbit brain to gums), A Recipe For Making A Paste To Prevent Gun-Barrels From Rusting, Whether Iron or Steel (combined bear's fat, badger's grease, snake's fat, almond, and indigo), prayers, and also religiomagical incantations such as one that will Spell-Bind A Thief So That He Cannot Stir. Derived from naturalists and primitives, seemingly de-evolving from the cohesive, complex, and better established medical beliefs of antiquity, powwowing in its varying success and accuracy in realms chemical, placebic, and psychic is not to be confused with those purer historical systems of combined practices far outliving it.

Precursors to modern medical remedies are attributed to, among others, the prehistoric Amerindians. Quinine was extracted from the bark of the cinchona tree and used to minimize malarial fever. The coca bush provided cocaine for the relief of both pain and the boredom of monotonous labor imposed by the Spaniards. A *Lechice ras* (Amerindian medicine man) confronted with afflictions of the respiratory track used *Eriodictyon glutinosum* growing profusely on foothills. When missionaries realized the efficacious value of this natural remedy they referred to the growth as *yerba santa,* "holy plant." The similarly esteemed cathartic *Rhamnus purshiana* of Southern California mountains became *cascara sagrada,* "sacred bark." While not a medication, a form of sutures as disturbing to us now as the thought of surgery without anesthesia is found in that which was used by generations of South American Indians employing the Sauba ant in the closing of wounds. The achievement of many medical goals has long been sought and attempted.

The Chinese, who attributed disease primarily to a disturbed equilibrium of yin (earthly) and yang (heavenly) forces, and who employed plastic surgery (for harelip) as early as A.D. 229–317, have given the world pharmacopoeia a vast quantity of drugs considered useless and sometimes even disgusting to Western senses. They have also, however, contributed important precursers to the modern pharmacopoeia. Cinnabar, the chief ore of mercury, cleared the nostrils for easier breathing, and also combined with other ingredients for the treatment of skin diseases. Goiter has long been remedied in China with seaweed, which contains iodide. The royal elixir, ginseng, is now popularly recognized far outside China although it has not yet found a place in conventional Western pharmacies. A less widely recognized origination is the global traveler that the Chinese, historically obsessed with qualities of vision, manufactured before Europeans had ever heard of them: spectacles, a name of dual implications, one of which describes a concept close to the basic concept of the microscope without which the history of genetics would not have existed in its present form.

Cerebral coordination of a body's nerve centers was anticipated as theory in the ancient Assyro-Babylonian god Ea. The Assyro-Babylonians, many of whose fantastic concepts blocked scientific progress for thousands of years, are responsible for the realization of a vast cornucopia of scientific facts in astronomy, mathematics, and various components of medicine, as is well known; these three areas of study form a triadic foundation to the whole of biological technology and medical history.

Antiquity did not profess a knowledge of the gene. It did contain many notions of heredity. These notions are the foundations out of which genetics, the science of heredity, was formed.

Contents

DISCOVERING DNA

Meditations on Genetics and a History of the Science

Section A
Discovering Genetics

1
History of Discovery

AGE OF SPECULATION

> The Philosopher's Stone transmutes one weight "but this transmutes another, and this again another, and so on, in so far that these mutations might be extended almost endlessly, just as one light kindles a second, and this second a third."—Paracelsus

In 1858 Charles Darwin began writing his *Origin of Species*. In 1953, James Watson and Francis Crick published their "Structural Implications of Deoxyribonucleicacid." In less than a century the question of the evolution of species had become a question of the creation of species, change had become a question of maintenance. The mechanism that controls individual changes and stability does so not only for the individual but also for the species, for nearly all species. It is the gene. The century between the theory of evolution and an understanding of DNA, the mechanism of heredity and thus of evolution, encompassed more than the length of an ordinary stone's throw. The quest for understanding even the basic premises of biological stability and change, a quest originating in the healing arts, was greater yet.

Stone: the cool surface is touched by two hands. One is the hand of modern man. The other is the hand of the now fossilized man called our ancestor. In heated discussions the eager speaker and the listener lean against the stone house, the brick building, the dwellings of our time paralleled to the cave, the flagstone patio parallel to the pebbled meadow where a single question is asked: What is life?

We may think of the two hands as the two strands with which DNA weaves a helical spiral linking past and future, old and new life. To discuss the two strands only in terms of time is to limit them unfairly. They encompass and express much more. Together, they are like the arms of the Great Mother. Current in the late twentieth century, the quest for the Father in spirit, God, is simultaneous with the quest for the Mother in matter, Earth. While cults are formed from yearnings, and create yearnings for the Father and Mother as World Parents described by various names, science also draws from within its fields. Within science dominant areas of the twentieth century include physics, search for the mysterious Father, and biology, search for the Mother.

"The great Mother was the head of the primitive family. She was priestess and sorceress in one. She was considered able to put pain to flight; she was the originator of the healing god. From the darkness of her womb sprang new life, the first miracle of which man became aware."[1]

The first miracle, then, of which mankind became aware was the springing of new life out of the "darkness of the womb." Yet the reproduction of our own species does not seem a likely candidate for the first awareness of miracles. Before reproduction man and woman encounter a multitude of surprising experiences and it is easier to imagine the taking of life or the transformation of a life, or even any one of the gifts to life such as light, dark, or rain as more likely for a first awareness of miracles. Life had to precede its reproduction. So, in the beginning there was life.

The first issue of manipulative awareness of life may be found in regard to health, in the maintenance of existing life. Secondly, how does new life, proceeding generations, and therefore existing life originate? Thirdly, how are these succeeding generations not only generally but also specifically similar to all other generations of their kind?

The evolution of thought has enabled us to perceive the importance of considering our origins when contemplating our health, to which heredity has become recognized as a vital factor in terms of the maintanence of health. The subject of genetics, however, considers not only individual but also special origins and even stirs the mind to consider our origins beyond species, in what we may or may not conclude is the primal soup from which the first cell emerged. The origin of life itself is a theme affecting our consciousness as much as the material process toward and maintenance of successive generations.

The marvelous interpretations of nature from which science has developed read like a great fantasia. In the contemplation of creation, the search for cause finds varying possibilities as to the specific element or elements from which life first sprang. Is the portrait of creation as an evolving primordial soup any more or less dramatic, romantic, or accurate than creation as the sudden emergence of all life directly out of Brahma's navel, or than water as the semen of nature, from which, Thales of Miletus proclaimed, all matter arose? The Ionian School of Philosophy founded by Thales (650–580 B.C.) was among those organized agitators of thought conceived before Darwin, Hippocrates, Plato, or Socrates. Not all of its members shared the views of Thales. The Ionian philosopher, Anaximander, enunciated the theory that humankind evolved from animals of a lower species, a view clearly anticipating Darwin. Wonder stares at the process of finding and losing an idea so capable of thunderbolting certainty into the realms of doubt which even yet embrace the question of our creation. Perhaps, between Anaximander and Darwin, the exploration afforded by scientific innocence is enough to justify the long silence of the theory of evolution. What silence or transformation awaits the resurgence of other recurring theories? Respectively, Anaximenes and Heraclitus, two principle Ionian philosophers, asserted that air and fire created life on this planet. Heraclitus may not have been all wrong in perceiving fire as his creator, for fire, among its symbolic interpretations, represents motion, and indeed a transitory motion. If evolution is correct in presuming that organisms change one into the other, even if this were to apply to only a few rather than to all of the life forms embraced by the theory, fire would not be an inappropriate symbol. In the case of the creator Air, Anaximenes combined science and religion by attributing intelligence to the atmosphere; this combining of approaches is of recurring significance, and is evident in our own times as culture and science strive to unite.

It is perhaps most important in the present day to stress that science does not threaten to deny the Bible of the creationist but attempts to understand it in its own language. As the psychic memory or sense of our origins seeks translation in the deeds of its creations, so are the languages of perspective discovered and translated between these members. Anthropology is psychiatry searching beneath the surface reality for clues to the origins of that surface.

Among the Ionian philosophers, Anaximenes was the only one approaching the later Platonic conception of the world, for it was Anaximenes who attributed intelligence to his creator, Air. Intelligence permits Idea, and to Plato the ultimate source of creation was Idea. This view has persisted into modern times. The present trend in physics has been likened to an attitude found at the beginning of this century when it was described as one in which "the universe begins to look more like a great thought than like a great machine" (Sir James Jeans).

There is frequently a reaction to approaching creation as the culmination of specific causes, including elements, which cries, "enough of this earth, water, air, fire! None of them alone could suffice to be responsible for so mammoth a project as earth and all of its life forms. Not even together is their force strong enough. It is one shared source from which we have sprung, a source subtler and stronger than this." Surely there is not only a common denominator but also a common source for all of life. Evolution explains the denominator in terms of the trunk of the tree from which branches of species emerge, while the originating seed of the tree is the cell.

No one alone of Plato's four elements, earth, water, air, or fire seems pervasive or powerful enough, and so we have Plato's god, Idea, from which these emerge. Modern science has yet to attribute plan to the primordial soup; there is nothing prohibiting the real possibility that, if we let them, all the missing or desired pieces, as contradictory and conflicting as they may appear and even be, will come into place much the same way, perhaps, as when creation let itself fall into place, if creation was and is a matter of letting.

The difference between creating and allowing for creation is a fine one. Reflection is as perplexed here as when faced with a perception such as Richard Dawkins's "selfish gene," where, as will be further discussed, we are to the gene exactly what we were just beginning to understand the gene is to us. The philosophers of antiquity gave no recorded vision as such of this gene, this house of nucleotides so apparently essential to the primordial soup. Not one of them had access to the astounding microbial worlds within the cellular pools explored on the dissecting tables of microscopes.

Fire, the perpetual alteration of opposites seen by Heraclitus as the Creator reminds the thinker of the duality running through many discussions concerning the definition of life. The two approaches to the question tend to separate at a certain point so that one deals specifically with the spiritual quality of life and one looks to a specific material origin of life. To encompass both approaches in one concept is apparently a difficult task. Was it the one original cell from which all life sprang out of the one God? We believe that spirit explains itself in religion, while matter explains itself in science, or that both are explained in either approach; the "or" implies a lack of consensus. Religion and the sciences attempt to satisfy our desire for a single explanation to embrace our whole world, including both matter and spirit. The attitudes of individuals, whether religiously or scientifically minded, ultimately divide our conception of heaven and earth. This division within humankind's search has resulted in a multitude of views we alternately doubt and cling to, thereby continuing the division. Today a new dawn is breaking. Even as new theories and religious groups or cults are blooming there is a desire for some unifying principle in which the war between the evolutionists and the creationists will be dissolved. Our whole vast and sublimated history sweeps in under the door to be fashioned according to technological as well as perceptual requirements. Which of our historical beliefs will be incorporated into the new outlook, or will they all; which, if any, of these beliefs will predominate? We could return to the Father from whom all blessings flow, or to the Mother Earth through whom they are shown. We could, conceivably, incorporate both, enjoy the awakening which recognizes the interaction of the World Parents described by Neumann in his *Origins and History of Consciousness.* And, I believe, we could perform this incorporation without slighting the theory of evolution or the Bible or any other spiritual doctrine or scientific theory, but the achievement of this goal is not yet at hand.

Perhaps our creation is simply the following picture, described in the combined language of science and religion, and beautifully expressed by William Irwin Thompson, who writes that "a bolt of lightning struck the waters, and the astonished molecules fell in chains to the quiet depths of the sea where they could whisper in the language of nucleotides to begin yet another chapter in the universal life of the Word."[2] Heaven and earth are beautifully and unusually reconciled in this statement. Yet too often it seems that only language lends a uniqueness to a statement that has recurred in one form or another for countless centuries. The wonderful thing about restatements is that if everything is forever saying the same thing but from perspectives that test their various vistas, we are learning new aspects of our subject. This reflects our need to be convinced again and again of traditional issues; in the course of testing one or being tested by one aspect of our world we get ourselves into terrible fixes which spur exploration into yet another aspect until we have tired of travel, feel that the whole has been viewed through its parts, and no longer challenge belief. The conditions of life are interdependent. The hope is that a resting place is envisioned for the future, a future that, in addition to distinguishing between them, realizes the sameness of all of our statements, and a new kind of activity capable of working with this knowledge. How far must we go into the realms of the abstract and the concrete before this is achieved?

"According to the Pythagorean mode of thinking, as reflected in the writings of Plato and his followers, Number itself existed before anything was created which could be numbered. 'Numbers are the sources of Form and Energy in the world' wrote Theon of Smyrna, the second century A.D. Platonist mathematician. The patterns of numerical relationships which occur in the processes of arithmetic and geometry were regarded as the formative influences behind all nature's productions."[3]

Four, the number of elements Plato's god drew out of chaos to create life, is a number of recurring and utmost significance to theories developed by both the scientist and the mystic. Hippocrates referred it to humors: blood, phlegm, yellow and black bile; these then corresponded to the four elements, as well as to four qualities, hot and dry, cold and dry, cold and moist, hot and moist. For Hippocrates life was bound up in these twelve aspects. Given that this kind of poetic assimilation is no longer fashionable, a correspondence of traditional elements to the molecular world has not, so far as I am aware, even been attempted. Table 1-1 shows the way in which Hippocrates be-

Table 1-1. The Correspondence of Aspects as Espoused by Hippocrates.

ELEMENTS	HUMORS	QUALITIES
Fire	Yellow bile	Hot and dry
Earth	Black bile	Cold and dry
Water	Phelgm	Cold and moist
Air	Blood	Hot and moist

lieved the elements, humors, and qualities to correspond to one another.

Hippocrates became the Great Physician by, among other things, addressing himself not to the origins but to the maintenance of life and to the susceptibility to and causes of disease. A susceptibility to disease similar to that of people today is recognized in apes and in prehistoric humankind. Notions as to the origins of these diseases, however, vary enormously not only in terms of language but also in terms of concept. Prehistoric man realized that a blow on the head from a stone would subdue an enemy; but what had the stone actually done? Perhaps the stone was simply the medium through which other forces were at work.

"Disease demons" struck a blow that made the very stone invisible. Four animistic forces traveled the air of the primitive, for, to the pre-Platonic mind, whether or not the air itself was intelligent, it was at the least a bearer of intelligence. Assault by any one of the four forces resulted in internal and invisible disorders. The forces fall into the following categorics, as described by Gordon.[4]

1. Heavenly bodies (gods, stars, ect.)
2. Aerial powers (demons, etc.)
3. Terrestrial forces (witchcraft, evil eye, etc.)
4. Subterranean forces (ghosts, worms, serpents, etc.)

Interestingly, the serpent (subterranean force) is also used to represent wisdom, cleansing and healing. Plato suggested that the most suitable physician is one who has suffered the ills to be treated. Are we, then, to see in the serpent that the cause of disease is provider of its cure? Inoculation restates what has been long known as a mystical declaration, describing two kinds of serpent, a positive and negative, potentially reconcilable polarities. Two serpents spiral around the staff of life to form the caduceus. The staff of the ancient herald is that which perhaps began, as Watkins believed, as the staff of Hermes, god of the trackways. The quality of the serpent, staff, spine, or path is also recognized in yoga. When we consider our physiological and psychic networking, the staff is readily seen, as has been shown elsewhere, to accompany not only the god of the trackways but also Mercury or the physician, Aesculapius. The caduceus remains the symbol of healing. Harmonized in the symbol are the otherwise conflicting positive and negative energies of life. In DNA we see the two serpents as the strands of the double helix: in self-replication of DNA, copying, these are the new and the old, visualized before and after replication as in a state of comparative harmony, a harmony resulting in the beginning and continuance of physical life. At the top of the caduceus is a pair of wings symbolizing "winged radiance," equilibrium. In discussing the recreation of the spiritual, rather than the physical self, in Tantric Yoga, C. G. Jung states that the "ajna cakra [a spiritual point or stage] is the mandala with the two petals, that looks like the winged seed of certain trees. It is quite purified from any earthly admixture, it has almost no substance, it is pure white light; so one gets the impression of something that really has taken to wings. And I think that this idea of the winged egg, or the homunculus in the second part of Faust, the artificial little man that flies about in his retort, is really the anticipation of such a possibility,—a man who has created himself again in new form."

Genetic engineering parallels the twentieth-century resurgence of a search for spiritual direction and revitalization, the remaking of the flesh parallel to the remaking of the spirit. The technology of the new biology holds many promises, not the least of which is medically oriented to existing life forms and to the potential problems of creatively conceived and new life forms.

Diagnosis, the first step of the physician, has changed considerably since antiquity. In antiquity diagnosis could take little consideration of heredity, unless, for instance, it was heredity as karma, an explanation exclusive of biological processes in their own right. It was comparatively late in the history of humankind and even rather late in context of the history of civilization that answers, of a physiological orientation, were formed to explain the resemblance between parents and children and we may be glad they so little affected the process of diagnosing inheritable disease. Various early Greeks, for instance, considered that the sex of the child was determined by which sex dominated in the sexual act; also that perhaps the characteristics of offspring are determined by the heat of the womb or from which testis the sperm came. George Louis Leclerc de Buffon (1707–1788), the great French naturalist, elaborated on the Greek thoughts with his belief that in the fruitful sexual act the male determines bodily extremities, the head, tail, and limbs of the child, while the female is responsible for the infant's internal constituents as well as for its overall size and shape. Because early diagnosis did not consider heredity as we know it does not mean that diagnosis was more general than specific. Medicine has passed through many stages. Only now do we have the ultimate physiological specificity, the cell.

Diagnosis employing a mystical concept of illness has been a favorite of many peoples throughout history. This form of diagnosis appears actually to dominate the history of medicine in early civilization and, to a lesser degree, has continued into the present. An appeal is traditionally made to a medicine man, whose function may rather be found in the shaman, perhaps the priest, the wizard, the doctor, or the exorcist. In this context it may be seen that not all doctors are fulfilling the prerequisites of their profession, as the mystical element seems to be missing. For those who would perceive the medical professionals in this way today, and in defense of not only the mystical element but perhaps its own evolution, we might consider that this element has become either so sublimated or so subtle that it operates unconsciously. Today the shaman is often sought within ones own self by persons practicing Christian Science or holistic methods of healing, including psychic healing. Both the physical and spiritual, are believed to require attention, and with both the patient is taking an increasingly active role. To perform this role the healer or self-healing patient reaches into the whole history of humanity's search for an explanation of who we are and how we may be changed.

The first healers of early civilization found it virtually impossible to separate spirituality from the material realm. Medicine of the early Babylonians, for instance, was completely in the domain of the priesthood. Priests were intermediaries, interpreters of divine wishes. They could appeal to the gods on behalf of a patient, and could expel the evil spirits causing disease. Chaldean wisdom was introduced to the

Assyro-Babylonian Empire by the conquering Persians. Priestly doctors, "wise men" (as in the "wise men of the East," New Testament), represented Chaldean wisdom in terms of its religion, magic, astrology (which permitted a correspondence between the parts of the human body and the signs of the zodiac), and divination. Powers responsible for disease were invisible and as external, in many cases, as the four animistic forces already discussed. It was not what man had but what had man that was the problem. The combating of disease, whether ill health was of material or nonmaterial origins, required assistance from both the material and nonmaterial realms. Prescriptions as well as charms and spells were used. Ailments were treated symptomatically. The Babylonians are noted for their symptomatic treatment and also for their recognition of the importance of blood. The ancient Babylonian myth of creation described blood as the foundation of life, for "man was created by mixing the blood of the decapitated god, Bel, with earth."

It is the elaborate Egyptian system of medicine that is credited with having provided the foundation for Greek medicine. And out of the latter realm emerged the Hippocratic oath retained, with the ancient caduceus, by the medical profession. In discussing the origins of the Hippocratic oath, the oath's direct relationship to Hippocrates has been disputed. The *Source Book of Medical History** states that "many believe the Oath was the pledge required of all who were admitted to [the] clan of Aesclepidae." And almost as possible as it is debatable is that "the Aesclepiadae were priests in the temples of Aesculapius." We are reminded of the differentiation experienced in more recent times by the body of society, the specialization which, in discoordination, separates every aspect of life even within each of the main branches of the material and the spiritual. Hippocrates in fact took no stand for or against the gods per se. However the Father of Medicine, Paracelsus, was clearly of a spiritual as well as a physical orientation which together spelled, for this great genius, metaphysicics. Paracelsus realized that the body is a chemical system. He blasted Hippocrates (among others), he reworked medieval alchemical theory, and he birthed the chemical treatment of physical illness, rejecting the prevailing Galenic four humor theory. (The only differ-

* *Source Book of Medical History,* Clendening, Logan, compiler, 1942, New York: Hoeber.

ences between Galen's theory of humors and that of Hippocrates, of whom Galen had been something of a follower, were that Galen cited mucus rather than phlegm, and according to Galenic theory blood was not just representative or corresponding to air but was a mixture of all four of the elements). In place of the Galenic theory, Paracelsus developed a chemical theory of humors as properties operating in the body.

"The physician who is also a philosopher, is godlike."—Claudius Galenus (Galen)

The Father of Medicine may have created a defense against the two men commonly credited with having been the greatest physicians of antiquity, Hippocrates and Galen; but with the latter, Paracelsus essentially shared a belief in preserving the union of philosophy with their art. It should also be noted about Galen that he promulgated the doctrine of "indication," lifting the art of the physician beyond personal experience, for the term comprised "whatever enables us to draw conclusions from as to treatment apart from experience." The interactions of philosophies emanating from individuals each confident of a particular view was, typically, shaping the future.

". . . there are two bodies, an eternal and a corporeal, enclosed in one. . . . Medicine acts upon the house by purging it, so that the spiritual body may be able to perfect its actions therein. . . ."—Aureolus Philippus Theophrastus Bombast (Paracelsus the Great)

A study of the history of medicine quickly makes apparent that however they are specifically adapted to their times the two aspects of healing, physical and spiritual, have worked in union. This union is worked out so that the parts act either simultaneously or consecutively, through the art of the physician or the attention of the priest. This union, expressed as a mutual exchange or a referral system among varying professionals, has never before recent centuries experienced so widespread and long-termed a disunion. Interestingly, now, in the same century in which fervent cries of the successes of a progress that has "outgrown" the past can be heard, there is an increasingly reputable adaptation of some of these very historical foundations.

The medical foundations that traveled on to Greece and to Rome evolved into the medical research and practice with which we are familiar today. They are incomplete, however, without the credit India shares with Egypt. This has become possible only since the archeological excavation of Sind which elucidated the remains of a civilization concealed for at least 6000 years. Sanskrit scholars have revealed that at an early period Aryans migrated from India. They established colonies in Egypt, Babylonia, Persia, and later in Greece and in Rome.

The Druids of Britain and the Scandinavians, states "Beal's Record," originated from Buddhist Brahman castes. Mathematics, geometry, gross anatomy, and medicine were cultivated by the Hindus to assist them in their religious worship, an important part of daily life. The Hindu system of medicine is known as Ayur-Veda, the science of life. It is said to have originated from Brahma and may be traced back to 6000 B.C. Hindu medicine may be divided into three periods. These are the Vedic, the Brahman, and the Buddhic. India's "scientific" medicine, like that of the Greeks, originated in mythology. Hindu tradition states that medicine and surgery were communicated by the supreme god Indra. The period of "scientific" or "rational" medicine in India did not begin until about 600 B.C., 5400 years after the date to which India's medicine has been traced. Ingenious methods for preparing items important to pharmacy and chemistry were contained in special texts.

Buddha was not concerned with the origin of creation but with the fact of it. An excerpt from the Vedic Hymns reveals another aspect of an Indian philosophy:

> "All gods are in the Divine Spirit; all worlds are in the Divine Spirit; and the Divine Spirit produces the connected series of acts which are performed by embodied souls. Him some adore as present in the element of fire; others as present in Manu,—lord of creatures; some as present in Indra; others as present in pure ether; and others as present in the most high Eternal Spirit. It is he who, pervading all beings in five elementary forms, causes them by the gradations of birth, growth, and dissolution to revolve in this world like the wheels of a car."

These wheels may be viewed from a different perspective as being each a cell, the spokes comprised of the genetic material, DNA. While neither the East Indians nor the Amerindians of the West perceived this material, the latter expressed an idea that anticipated knowledge of the bacterial origin of disease.

The quality of the Amerindian approach to physical life and to medicine is immediately evident in the fact that birthing is not just midwifery, it is spiritual midwifery, for a soul is being born. Among early American Indians three important techniques for combating disease are known to have existed. The medicine man was a frequently sought resource, but personally directed sweating or isolation were also called upon. The Aztecs are said to have sometimes surrounded the patient isolated on a mountaintop with poles on which animals were hung. This served a dual function. It warded off evil spirits and it suggested bodies to which the patient's disease might be transferred. Indian tribes frequently believed that a disease or affliction, in terms of either its cause or its symptoms, or both, could be transferred between persons or from a person to an animal. How endless this seems: a domino effect of persons contacted and vacated by unpleasantries apparently doing nothing more than seeking a willing victim. Although transference is said by Gordon to exist in comparatively modern times in the East, much as leeching is reputedly still in limited practice in parts of France, this is not only unsatisfactory to moral consciousness, it is readily perceived as not really taking care of the problem. The latter recognition, that the problem is not adequately solved by transferance, is current not only among medical practitioners of the classic mode, but also among psychic healers who desire to "dissolve" an ailment lest, as in a crowd before a healer, one person reports the sudden absence of a headache while another person simultaneously is discovered to have suddenly acquired a headache.

Perhaps indeed the victim of disease, in addition to being vulnerable, unconsciously welcomes life's threats or discomforts through some distortion of need. Disease was believed by the Iroquois to result from the unsatisfied craving to fulfill an inborn desire. With this belief the Iroquois have been credited with anticipating psychotherapy. More to our purposes is that an anticipation of the bacterial origin of disease is acknowledged in the Myettas' belief that flying insects are responsible for disease, a version of the concept that disease is airborne (disease demons). Proponents of genetic engineering may say that the bacterial disease culprits should make good with us, which is what they seem to be doing as bacterial "factories" manufacturing the medical promises of a genetically engineered future.

Efforts to understand and to combat disease reach far into antiquity, beyond hundreds and into thousands of years ago. Cro-Magnons are believed from anthropological data to have exerted determined efforts to care for their diseased and injured. Even if we should discover that such primitive individuals are not in fact part of our own evolutionary tree, the fact that humanlike creatures exerted this effort, even if they were not human beings, is of interest for the history of medicine. Whatever threats may be perceived as imminent from the technologies of molecular biology, the medical applications of these technologies strike a sympathetic cord deeper than was perhaps imagined to exist by those who have sought to, at the very least, slow the ever-increasing enthusiasm for well-being in order to ensure that our efforts do indeed culminate in its encouragement and not in that reversal which a blinded progress could deliver.

When disease was finally attributed to external and invisible elements a great contribution was made to the future of medicine. Medicine is once again in its infancy, this time, however, in view of the attention it pays to the internal and visible. For the lay healer this internal and visible element is the psychic "star". For the biologist it is the cell; deeper yet it is DNA.

AGE OF OBSERVATION

To what strange interiors do we look for a reflection of ourselves? Who has seen homunculus? If, in physical terms, we were to return to our origins, how strange it would be to have no name for the destination of our path, no visualization of its shape or size or place.

"Omnis cellula e cellula."[5]

Exteriors but alternately consume our observation. Inevitably there is a return to a reflection upon, or observation of, interiors. With each return to interiors perception seems to be more intense, to reach deeper terrains. Finally, the depth is right for a sounding. Now the sound is "cell."

"Cells" were described in 1835 by Johannes Müller. Avogadro had already conceptualized the molecule. Cell theory was established in 1838 by M. J. Schleiden and in 1839 by Theodor Schwann, giving biology an atomic basis. Schleiden, one of the first scientists to perceive the cell as the founding point of any life, also was one of the last to proclaim that plant sexuality does not exist. Without intention, he kept finding himself in a presexual realm, the undifferentiated sea out of which specificity emerges. Schleiden does not appear to have been the kind of man who would have interpreted his life philosophically to any great extent. Moreover, although science may be an act of questioning, Schleiden seems to have been inclined to a frustration with the uncertain. So it is that his tenacity on the matter of plant sexuality required "a whole team of cytologists . . . to work for years," as Dunn puts the problem in stylistically understated fashion,[6] until in 1856 Schleiden was utterly disproved on the matter of plant sexuality by one of his own students. In terms, however, of his other work, in cell theory, the effect was just the opposite.

Cell theory had not been convincingly put forward until the work of the German microscopists Schleiden and Schwann. The theory proposes all plants and animals to be constructed from small fundamental units. These units are cells. The theory is one that was developed from observations made of higher organisms. Even though differentiation in lower and higher life forms has strikingly different results, cell theory holds for simpler forms of life including protozoa and bacteria.

Genetics has had a significant interrelationship with physics. Brownian movement, elucidated by Robert Brown, who also discovered the nucleus of the cell as seen in the epidermis of orchids, provides an example of an intertwining of the languages and perspectives of physics and biology. Brownian movement, "is the chief of those fundamental phenomena which the biologists have contributed, or helped to contribute, to the science of physics,"[7] and is therefore a kind of reverse of what happened when Schrödinger inspired physicists, to contribute to biology. The quivering motion in Brownian movement was early thought to result from calorical changes or from an obscure chemical action between solid particles. Later, however, Christian Wiener ascribed it to molecular movements, and "it is indeed due to the impact or bombardment of molecules upon a body (and this) bombardment causes our particles to behave just like molecules of unusual size. . . . The Brownian movement was hailed by some as supporting Liebniz's theory of Monads, a theory once so deeply rooted and so widely believed that even under Schwann's cell theory Johannes Müller and Henle spoke of the cells as "organische Monaden.'"[8]

While Brownian movement was erroneously suggested by George Johnston Stoney, who conceptualized the electron, to be a source of bacterial energy, cell theory was fundamental to the developments of molecular biology and genetics.

Although we consider natural human procreation to require a sexual act now understood to result in the union of two cells, the realization that life grows from this one unified cell began to emerge at a time when sexuality was scandalous and many of the facets of life a dubious reality. But facets are fundamental to wholes. When Schleiden stated that recognition of the "existence of an essential life in each separate cell" is fundamental to a study of the whole, he campaigned for a recognition that has exceeded his definition or speculation about the life within a cell; genetics describes a life so full that its potential is matched only by wild fantasy.

The twentieth-century outlook on genetics would be impossible without the role performed by bacteria in research. Bacteria are, in the words of S. J. Gould, natural precision designers with built-in magnets; they reveal "biology's meticulous engineering." The first bacteriologist was the Dutch microscopist Anton van Leeuwenhoek. He lived a quiet life in Delft, grinding his own lenses for microscopes that he himself constructed. The microscopes described by Descartes earlier in the century were not powerful enough for the visualization of bacteria, but van Leeuwenhoek's were. In the late 1600s he began publishing in the *Philosophical Transactions of the Royal Society* descriptions of the remarkable world he had discovered. The new world excited great interest but enthusiasm was lacking. It slept in the problem of investigation. Intellectual adventurism was frustrated by the states of chemistry and medicine, which were still striving toward professional maturity. Even had the scientific environment been sufficient to prompt the right questions, progress would have been stymied until the methods for obtaining pure cultures had been developed. In a sense, van Leeuwenhoek's work was premature in terms of readiness for exploration; yet it was also timely in that it fed curiosity.

The 1600s were a time in which interiors were reexamined. The microscopic depths of organisms were reexamined by van Leeuwenhoek when he perceived microscopic life forms. And the era saw Boyle show that Aristotle's four elements could only be considered as symbolic. Boyle, who laid the way for major developments in chemistry, was accused of undermining official religion, expounding the view that scientifc investigation worked for, not against, his own potential to be a "good Christian." In the same period Rembrandt reexamined architectural and human interiors with paint, perceiving dark and light in new ways.

The martyring of Giordano Bruno in the year 1600 initiated the first century of modern science, though Bruno suffered not for a scientific cause but for "free imaginative speculation." It was as a defender against the oppression then severely characteristic of the Catholic church, that his name was given to the newborns of hundreds of Italian families.

The doctrine of the microcosm came into great favor among philosophers of the Renaissance like Bruno and also Paracelsus. In Bruno's execution, writes Alfred North Whitehead in his *Science and the Modern World,* "there was an unconscious symbolism: for the subsequent tone of scientific thought has contained distrust of his type of general speculativeness." Galen's form of speculative diagnosis in medicine, a concept rejected by the Empirics and ignored by the grossly limited Methodists, even when attached to the practical confinement of experience, had seen itself as a concept of indication rise again in Bruno only to face again the overpowering social determination to refuse an education in the subtle. Meanwhile, Newton had been born (1642), Galileo had died (1641), Shakespeare was writing, Bacon was publishing, and Descartes' *Meditationes* and *Principia Philosophiae* had just come out. Science was preparing for a new mathematics by which to become more fully equipped, and genius was drawing clear distinctions throughout what had been sadly perceived as a muddled stage in the evolution of thought.

The scientific community of the 1600s perceived the organic world in terms of ascension. Pierre Teilhard de Chardin has described this attitude in terms of humanity as "the ascending arrow of the great biological synthesis." Ascension explained the ultimate perfection of humankind. The tremendous difference between ideas of perfection and the reality of the imperfect denied the attributing of many characteristics to life forms believed to be undeserving of them. Plant life was hardly life at all. Plants could not be expected to share the characteristics of a more highly ascended life form. Insensitive and asexual, plants were at the bottom of the "chain of being."

The characterization of plants as lowly is a curious and isolated fact of history. Plants are vegetation, vegetation is food and has been anthropormophized throughout previous ages. Vegetation has provided sacred symbols. Throughout mythology heroes such as Osiris represent vegetation. Even Christ, as described by Neumann, can be seen as a symbol of sacred plant life, having been born in the "house of Bread," which is the translation of Bethlehem. Plants have also been significant medicinally, aiding the well-being of and saving countless lives for many centuries. The meagerness of plants in size and substance, allowing them to be stepped on without very great effect in ages accustomed to forests or ignorant of their value, as well as the mysterious nature of plant life, contribute to a devaluing characterization. Furthermore, organisms more closely resembling, if not attaining, the highly ascended humans have naturally been of greater interest and immediate potential.

What right has the lowly plant? What example does it set the pious naturalist? The answer in seventeenth-century terms is that the plant has very few rights, and whatever example it sets, the pious naturalist must reaffirm what that naturalist already takes to be true. Poor plant? No, rather pity, if pity you must, the tormented world that considers the possibility that plants might have sexuality as completely dangerous to thought in general. That this sexuality might go so far as to express itself in male dominance would be and was the second shock. The stage was set when Casalpino (1583) discarded a plant classification system based on variables, roots, stems, leaves, etc.

Whoredom in the long-forgotton mythology of the Great Mother was more often than not both symbolic and sacred. In plants the whoring implied by male dominance was literal and awful. In his *Origins of Mendelism,* Robert Olby illustrates the response to plant sexuality. He quotes J. G. Siegesbeck, Professor of Botany at St. Petersburg, who exclaimed, "what man will ever believe that God almighty should have introduced such confusion, or rather such shameful whoredom, for the propagation of the reign of plants. Who will instruct young students in such a voluptuous system without scandal?" The demonstration inspiring this exclamation had come in 1694 from Rudolph Camerarius, Professor of Natural History at Tübingen.

Camerarius provided the first experimental proof of plant sexuality, as described in his *De Sex Plantarum Epistola.* Actually, plant sexuality had already been recognized centuries earlier by Tyrtanius of Lesbian-Eresus (Theophrastus), the very Father of Botany. Uncovering the veil that had not been lifted in the scientific world since antiquity, Camerarius rediscovered the male/female elements of plants and the essentialness of their union for the formation of new individuals. Plants, with animals and humans, shared a common denominator, their sexuality. No pious naturalist could rightly deny this, however upsetting it may be. A botanical rule was established. The virtually inevitable exception to the rule appeared in the female isolated from males of her own species. She still produced seeds. Camerarius suspected that foreign pollen was at work here. Hybridization became an attractive problem in the midst of a shocking shift in thought.

Between the rediscovery of plant sexuality and the first scientifically acceptable plant hybrid looms the prestigious and arrogant apostle of plant sexuality, Carl Linnaeus. The Swedish botanist had determined and proceeded to teach the new doctrine of plant sexuality to young students at Uppsala. He also dared to create a system of plant classification of which the sexual organs were the very basis. The results of these activities included a magnified awareness of the redefined botany for science, and for Linnaeus a personal self-aggrandizement. Olby describes the eighteenth-century botanist as a man who "discovered" himself to be a kind of avatar within the domain of science. Indeed, the present-day names of plants have their origins in the work of Linnaeus. His system of classification, however, was unbalanced to a biological, if not to an Aristotelian, point of view (accentuating sexual to the exclusion of other characteristics). An unweighted system of classification, writes Patrick Collard, a system in which all characteristics are considered, was developed by Michael Adamson and published in 1763 as *Familles des Plantes.* Until the twentieth century this work remained virtually unknown. The unweighted system of plant classification in its rediscovery has been acclaimed the most suitable of any known system of plant classification for use with modern computer technology. Computerized and Adansonian taxonomy have become synonymous. However unbalanced Linnaeus may have been in preparing his system of plant classification, it was he who zealously dared reply in essay form to the competition announced by the Imperial Academy of Sciences, St. Petersburg, in 1760. Competitors could either defend or refute the attribute of sexuality in plants. Linnaeus's essay, arguing for plant sexuality, won him the prize of 50 ducats. But as yet, no plant hybrid had itself been experimentally produced. In the search for natural hybrids Linnaeus, with an overanxious yet lazy eye, concluded that any plant expressing the characteristics intermediate between two other species was not just similar to the other two species and was not merely unique in its combination, rather, it was a hybrid. He considered many plants to be hybrids which were not hybrids at all.

During the year 1760, in which Linnaeus won the academy's prize for a defense of plant sexuality, the first viable experimental hybrid was in the process of being born with the assistance of J. G. Kolreuter.

"If there is one person to be chosen as the direct scientific ancestor of Mendel, de Vries and the other founders of genetics, he must be Kolreuter, for it was Kolreuter who, in 1760, in a deliberately planned program produced the first plant hybrid resulting from an experiment."—L. C. Dunn[9]

Experimental genetics, then, began with two tobacco plants, *Nicotiana paniculata* × *Nicotiana rustica,* deliberately and carefully hybridized by a young Joseph Gottlieb Kolreuter. Before the following century had barely passed its halfway mark the observations of Pasteur and Charles Darwin had also contributed knowledge that was to be cultivated and refined by many future generations. But a history of the science of these past centuries makes them seem conceptually awkward at best.

Linnaeus and Kolreuter may be seen as extensions or mirror images of each other, polar images; the former being overzealous and almost amateurishly studious, the latter being appropriately studious, but lacking perhaps psychological stamina. Both the importance and accuracy of Kolreuter's work were criticized by those seemingly immovable forces who yet denied the existence of plant sexuality. However,

Kolreuter was successfully defended by several scientists, including his former neighbor, Carl von Gaertner. Gaertner published widely on the subject of hybrids; he repeated and extended Kolreuter's work, and he proved to be, beside Kolreuter, a substantial benefactor to the Mendelian future. He knew well the work of Joseph Kolreuter both on page and in garden soil. Kolreuter had been known to plant the seeds of his plants wherever he could. They grew in pots along the windowsills of friends' homes, in pots carried in his two hands during his various travels, and in the garden of the Gaertner home in Calw. While Linnaeus produced what he called hybrids, doing so in an environment the scientific community did not recognize as the atmosphere of controlled experiment, the work of Kolreuter resulted in the first scientifically acceptable hybridization, despite the seemingly desparate means by which he sought to nurture his seedlings.

When Kolreuter's circumstances finally improved to the extent that he had an entire botanical garden and a staff of gardeners at his disposal, the improvement was the result of his being selected, over the uninterested Linnaeus, for a professorship at Karlsruhe. As Professor of Natural History and Director of the Margrave's gardens, Kolreuter's prowess as a pioneer was challenged so severely that his experimental work succumbed to the flagrant cruelty of his own head gardener, who choked the garden plants with weeds and destroyed the remaining greenhouse plants with lack of heat.

The paradise Kolreuter had finally found began to collapse around him. The head gardener became an archvillain and finally the relationship subsided in Kolreuter's relinquishing the second portion of his title, Director of the Gardens, to Carl Gomelin. Kolreuter the pioneer did not reappear. In spite of the modest controversy his early work had caused, he was not widely recognized. This lack of recognition changed when Charles Darwin began his study "On the Origin of Species." For this study special consideration was made of both Kolreuter's and Gaertner's publications. With frequent reference to their work, Darwin made known the pre-Mendelian foundation of genetics.

During the middle 1700s and 1800s important findings were made. Louis Pasteur's work in fermentation in the middle 1800s has certainly had resounding effects. Fermentation performs an important role not only in the history of science but also in the history of organisms and the cell. "Having arisen in an oceanic broth of organic molecules, [the primal organism's] only recourse is to live upon them. There is only one way of doing that in the absence of oxygen. It is called fermentation: the process by which organisms derive energy by breaking organic molecules and rearranging their parts."[10]

But what is the structure of those organisms deriving their energy in this way, and how is it maintained? A study of anatomical structure and function externalizes the fundamental questions recurring throughout the ages. We recognize a like system of procreation in man and beast, rediscover it in vegetation. We even try to manipulate this system and sometimes we succeed. But we are looking at large bodies: knowing that their movement is purposeful, and even knowing a certain amount of what that purpose is, does not tell us about the minute details propelling it. For the reason that the latter had yet to occur, botany, biology, physiology, and anatomy shared for a long time nothing but the germ of an idea with either Darwinian evolution (1859) or the surgeons' text, *Anatomy, Descriptive and Surgical* (1858) which has since become known as *Gray's Anatomy,* nor certainly with D'Arcy Thompson's eloquent and again structural *On Growth and Form* (1917).

Doubtless even the psychic who perceived anatomical structure and function for centuries before the very existence of formalized science had little or no inkling of the deeper chemical meaning of life, alchemy aside.

Numerous and important developments were taking place in the structuring of the future. Among these was Pasteur's introduction of the "germ theory of disease." This occurred in 1865, the same year in which Mendel quietly read out his first paper, later known as the paper of 1866, the year it was placed into the archives.

The path to taking the mystery out of simple physiological reproduction has been a long one. In the course of this path the literal and symbolic overlap, and one seems to be confused with the other. Grecian theory had described a pervasive sperm secreted from not one but all parts of the body during intercourse. Animated by an unknown force, sperm emerging from each part of the body subsequently reproduced that part of the body from which it had been secreted. This train of thought, as we recall with some amusement, led to the idea of a fully formed embryo, in miniature, in each sperm cell. Our amusement is founded on a confusion of the revelatory nature of the literal and symbolic. Each is quite natural to reality, and the literal and symbolic actually coexist, and rather insistently. In situations like the Greek theory we see how they do not overlap as acceptable perceptions of a single reality—homunculus, for instance, is fine symbolically but not literally; in other situations we may see how they do overlap. The Greeks were not alone in the idea of a homunculus. Nor, certainly, was theirs the only medical language in the world. It had developed, typically, from and to other languages.

"Now pangenesis is really a version of the ancient Hippocratic doctrine of the body tissues which was restated by numerous authors after Hippocrates."—Robert Olby

The theory of pangenesis was a more reasonable, if still unsuccessful, attempt to solve an internal mystery. In pangenesis cells throw off pangenes. The protoplasm constituting pangenes was quite correctly, considered important, even vital. Its contents include not only the long-recognized protein but also the incredible mitochondria of recent investigation.

Pangenesis was on a track later clarified and held legitimate even though the picture given by pangenesis had been naive. This is a characteristic familiar to farfetched notions, and, in terms of cultural symbolism and keeping heredity under consideration, farther fetched notions naturally come to mind. Such a notion is evident in pagan rites performed to fertilize the land. Among certain peoples furrows were dug into which, in historical times, men temporarily fitted themselves, there to merge the sweat of copulation with the moisture or dryness of the earth. Research into psychology may validate this practice, and research into psychic phenomenon may yet further support it, at least to a degree. If the chain of thought within one's mind is important to the development of the fetus growing within the pregnant mother, it can be as important for the well-being of the farmer and the fertility of the land whose fruits will nourish the child as they have its mother, and the rite here described simply an exaggeration of a valid principle which may be much more simply and what we would now deem appropriately expressed.

It is not discrediting for one man to extend views only some of which we find to be the least bit acceptable at a given time, with an edited interpretation over a period of and perhaps throughout time. Darwin's theory of evolution seems less extravagant symbolization and literalization. We are told, quite accurately in terms of science, that in Darwin's day "nothing whatever was known of the fact of heredity" (Sir Gavin de Beer). Braving the void, Charles Darwin set out to show that variation is inheritable.

"Ever since its birth in the Ionian Islands almost three thousand years ago, Western philosophy has been divided between two seemingly opposed attitudes. According to one of them the authentic and ultimate truth of the world can reside only in perfectly immutable forms by essence unvarying. According to the other, the only real truth resides in flux and evolution."[11]

This quote presents us with a great misconception made possible within society by the misaligned condition of quite reconcilable opposites. Can it really be true that the science and nonsecular philosophy that have become trusted foundations for an objective future are not ready to relinquish the fruits of a specialization that has served not infrequently to keep these two separate? Surely "continuity in change"* or "the principle of permanence within change"† are legitimate concepts. Form is mutable; essence is not mutable, nor is essential form. The conditions by which life forms exist on this planet will accommodate the variation of form only within the context of limiting prerequisites. Life is both varying and unvarying, nurtured and natured, perfect and imperfect. Whether or not life in its various forms originated by way of an evolution or in the spontaneous appearance of species is argued not only by anti-Darwinists. Argument has surrounded Darwinism since its formation.

"Evolution is something that happens to populations, and without a mathematical theory, connecting the phenomena in populations with those individuals, there could be no very clear thinking on the subject."—Sewall Wright

Appearing nearly half a century before the rediscovery of Mendel and long before the highly statistical population genetics, the evolu-

*Anshen, Ruth Nanda, in her introduction to *Across the Frontiers,* Werner Heisenberg, Harper & Row, 1974

†*Physics and the Alchemy of Consciousness,* Pir Vihyat Inayat Khan, Sufi Order Publications, 1978

tionary theory of Charles Darwin is not so much a theory of heredity as it is a theory about what heredity can do. Genetics is a study of both the mechanism of heredity and its achievements on both the short-term (individuals or several generations of individuals) and long-term (evolutionary) scales. Scientists concerned with embryology and development, such as Bateson, Castle, and Morgan, who had become converts to genetics, found that developmental problems were displaced with older questions concerning the mechanism of evolution. The basic measurement of evolution is found in gene frequency, as a parameter of natural selection. It was stated as such by Sewall Wright in his 1931 paper:

> "Evolution as a process of cumulative change depends on a proper balance of the conditions, which, at each level of organization—gene, chromosome, cell, individual, local race—make for genetic homogeneity or genetic heterogeneity of the species. While the basic factor of change—the infrequent, fortuitous, usually more or less injurious gene mutations, in themselves, appear to furnish an inadequate basis for evolution, the mechanism of cell divisions, with its occasional aberrations, and of nuclear fusion (at fertilization) followed at some time by reduction make it possible for a relatively small number of not too injurious mutations to provide an extensive field of actual variations. The type and rate of evolution in such a system depend on the balance among the evolutionary pressures considered here."

It is no sleight-of-hand that allows the gene, which provides the "basic measurement of evolution," also to be "an inadequate basis for evolution." These statements emerge, as does the relationship of studies in genetics and evolution, out of studies in population genetics. Population genetics has been described as the subject at the very core of general biology. There is much to recommend it. The study has reunited such fields of biology as genetics, ecology, paleontology, and systematics. Furthermore, while numerous questions in genetics have been given or can be seen to eventually be given new form by chemistry and physics, population genetics remains pure, and before the emergence of new evolutionary theories was thought "likely to stand for many years to come as the exemplar of the original purposes and methods of genetics" (L. C. Dunn).

With what is referred to as the great evolutionary synthesis of the

1930s and 1940s population genetics shared a vital role. Population genetics evolved out of a desire to harmonize Mendelian ratios within and outside the laboratory. "At the beginning of this [twentieth] century there were various misconceptions about the relationship of Mendelian ratios found by genetic analysis and the actual proportions of different phenotypes and genotypes observed in natural populations."[12] The study took on a sophisticated mathematical nature. The greater role of the gene was as yet undetermined; researchers involved in population genetics at its outset strove simply to determine whether or not Mendelian genes were the factors responsible for evolutionary biological variability. Out of this study a picture of the gene pool emerged, the collective of what had been previously studied in terms of individuals. The tributaries of that pool, the three problems upon which the study focuses, depended for advances in their solving upon rigorous mathematical theory and analytical methods in sustained application to the following areas as they are stated by Avers.[13]

1. Genetical correlation among relatives of an interbreeding population
2. Analysis: mating systems (inbreeding, assortive and random mating, etc.)
3. Analysis: evolutionary forces (selection, mutation, migration, chance)

That we can allow for both selection as well as random genetic drift, chance, is an after-the-fact prerequisite to understanding natural mutation which, broadly, is the condition necessary for much evolutionary change. Approaching metamorphosis, mutation, as a form of evolution is necessary not only to evolutionary change of a biological nature but also to any change, emotional, spiritual, or psychological, social, cultural, etc., and at every level of existence we can see that the history of genetics is an unfolding of more generally applicable principles. Therefore, the fact that mutation allows for both selection and chance in the changes that maintain life and its diversity is more profound than may at first appear. But the concept of chance can be alarming. In a slim and widely treasured volume, *Chance and Necessity,* Jacques Monod leaps in and out of disconcerting stances such as the following. Having discussed various forms of mutation and the fact that these particular events are called accidental, Monod goes on to say that "pure chance, absolutely free but blind, [is] at the very root of the stupendous edifice of evolution: this central concept of modern biology is no longer one among other possible or even conceivable hypotheses. It is today the sole conceivable hypothesis, the only one that squares with observed and tested fact. . . . There is no scientific concept, in any of the sciences, more destructive of anthropocentrism than this one, and no other so rouses an instinctive protest from the intensely teleonomic creatures that we are. For every vitalist or animist ideaology it is therefore the concept or rather the specter to be exorcised at all costs."

An interrelationship between chance and necessity becomes inevitably allowed for by the gene. Necessity can arise by virtue of modulated chance causing modulated change. This is an "and vise-versa" situation symbolized in the sweep of curves on the undulating surface structure of the double helix.

Evolution has traditionally been thought of in a similar way: although the line of progress in evolution has been depicted as a straight one it is a line composed of invisible curves (punctuated equilibrium, a new version of the theory of evolution, makes these curves into sharp angles). Studies in evolution and genetics formed their first mutual bond in population genetics. Theories concerned with heredity in populations naturally acquired a new start when heredity was linked to the gene by virtue of the concept of hereditary units. Wilhelm Weinberg is noted for having provided the most comprehensive early treatment (1908) to Mendelian theory applied to randomly mating populations. While Weinberg's work did not have an immediate effect, it dealt with gene frequencies, interactions, and environmental variance, establishing a theme that had to be recognized by independent studies before the work of Weinberg and the study of population genetics could be fully realized. The first period of rapid development in this area began between 1910 and 1920. Population methods were used to discriminate between genetical hypotheses. Sewall Wright's seminal paper of 1931, "Evolution in Mendelian Populations," provided one of the mathematical models for the operation of selection. This paper, along with Fisher's book, *Genetical Theory of Natural Selection* (1930), and J. B. S. Haldane's book, *The Causes of Evolution* (1932), as well as work by the Russian S. S.

Chetverikov (1926; English translation 1961), spawned population genetics as a recognizable biological field.

The study of inheritence is one which focuses on differentiation. Yet, having conceptually differentiated the physical characters of numerous life forms, "character by character," so that an understanding of their similarity and dissimilarity may be attained, we find that "characters which we have differentiated insist on integrating themselves again; and aspects of the organism are seen to be conjoined which only our mental analysis had put asunder. The coordinate diagram throws into relief the integral solidarity of the organism, and enables us to see how simple a certain kind of correlation is which had been apt to seem a subtle and a complex thing."[14] Is it not reality but instead thought which is complex?

Thought is often stymied by the complexity brought to bear by new perspectives; new perspectives have a tendency to seem threatening to existing ones. The view of certain primitive mythologies that man was created prior to beasts is as little jeopardized as modern evolution theories by the discovery of a genetic similarity in humans and apes. The similarity is of an at least 98 percent chromosomal matching between humans and chimpanzees, the chromosome number 3 pair being clearly identical. "Such a remarkable degree of similarity makes difficult a precise explanation of the large biological difference observed between these two closely related species."[15] It also provides the basis for hopes that future work in this direction will help elucidate the nature of the origin of physiological differences between species.

How is the recurrence, typical of many scientific theories, applicable to the future of the idea put forward by Diogenes that "the variation in the souls of brutes, which differ from men and from one another in intelligence, is only due to the air in its various conditions of moisture and warmth." An appropriate vision might exclude biology in the traditional sense and encourage philosophy to tackle the nature of variation by virtue of a symbolistic approach to Diogenes' interpretation. Such exclusion is not necessary for Aristotle, who, as the "first biologist of antiquity to maintain that there was a gradual succession of living beings from the very simplest to the most perfect form," seems at first glance to have already been resurrected by Darwinism.

Darwin reinitiated a global interest in the integrating factors of disparate forms, an accomplishment that reverberated far beyond its source. For this he should be blessed. The broader implications of Darwinism in this light are possible only because of the scientifically unique approach that brought evolutionary theory into wide visibility. This "uniqueness does not reside in his support for the idea of evolution—scores of scientists had preceded him in this. His special contribution rests upon his documentation and upon the novel character of his theory about how evolution operates. . . . The theory of natural selection is a creative transfer to biology of Adam Smith's basic argument for a rational economy: the balance and order of nature does not arise [by direct gratuity] from a higher external (divine) control or from the existence of laws operating directly upon the whole, but from struggle among individuals, for their own benefits. . . ."[16]

At first glance an article in *Science,* November 21, 1980, entitled "Evolutionary Theory Under Fire," might seem to be an open door to antagonism of the creationists. It is not. "No active geneticist, embryologist, systematist, anatomist, or paleontologist doubts that life has evolved. What such biologists do argue about is *how* life has evolved" (Niles Eldredge, American Museum of Natural History). Evolutionary theory is undergoing various changes. The article referred to challanges the current rendition of Darwin's explanation of our origins. Evolutionary theory is still alive but it is undergoing revisions. This will perhaps suit creationism more than it knows. Creationism in its current form appears highly desirable of reestablishing science and creationism as quite divergent fields by the sheer force and tactics it employs in an effort to achieve just the opposite; this is occurring when both science and creationism are on the brink of naturally culminating an initiation which recognizes their mutual search, a time when they might discover their cohesiveness and also to redefine their individualities within that oneness. The framework of evolutionary theory will doubtless withstand the current criticisms and in so doing will be reaffirmed.

Science is concerned with microevolution and macroevolution. Microevolution categorizes changes within a population, while macroevolution tackles the origin of species and their evolutionary relationships. Darwinism is not only expanded, it is redefined, if inconclusively. As theories develop it is important to remember that a theory is never proved, only disproved, as we have been reminded in the wake of the modern version of the Scopes monkey trial which punctuates either end of the 20th century.

Neither evolution nor creationism can be proved by laboratory experiment. They can be and are both perhaps literally or symbolically evidenced in the laboratory and in field work, but they are never proved in a scientific sense. It is more realistic than mystic or pessimistic to say that one does as much as the other: However many theories in science and creationism, science and creationism each functions not only as a liberator but also to reassure us—inconclusively in terms of experimentally provable origins and originator. Was Paracelsus wrong, then, when he stated:

> "How, then, can it be that we should withdraw from this noble art of medicine, or from philosophy itself, when we see so clearly their force and power, which alone confirm us, and deservedly so, that we should place the greatest confidence in them? For we have not applied our minds to believe, learn, or imitate things which cannot be proved and attested by most true and certain evidence conjoined with reason. When Christ hung upon the Cross, if the sun and moon had not been affected with a kind of sympathy, so that they were deprived of their light and darkened, and had not the earth been shaken with a fearful tremor, and had not other signs been manifested at His nativity, no one would now believe in Him. They naturally teach us to see and know this, that Jesus Christ is God, and took human nature upon Himself . . . they cause and even compel us to believe in them, so that we should not withdraw from them even up to the time of death. . . ."[17]

However physical reality is interpreted there remains the concept found among both symbolists and literalists that a pervasive force operates throughout this realm, and to some thinkers the physical systems through which this force operates are viewed as divine. The twentieth-century controversy in this area provides an opportunity for clarifying and redefining many issues that need not antagonize one another, nor perhaps even be irreconcilably contradictory.

The analytical psychologist, Erich Neumann, depicts contradiction as not only acceptable but preferable. Between every two contradictions is a potential harmonizing factor that otherwise might not have been tested. Fighting against this potential harmony further separates the two views, and on their own they flounder as parts irrespective of their potential whole. If we can accept in their own rights the concepts of both inconclusion and conclusion, the concrete and the fluid, the theoretical striving of both science and creationism, their respective identities and interrelationship may become so clear that we can work with them simultaneously without controversy. In this way the scientist naturally and automatically translates and incorporates the concepts of creationism and philosophy, as physics is tending with the idea of Idea, and as creationism might do with science (without, necessarily, making a science of creationism), for they express each other. To categorically deny the validity of either is to deny life as a whole, and life as a whole is what each seeks to explain. To categorically deny independence to either is to imprison both.

There are four systems that could be used collectively to describe life on earth. There are the physical and nonphysical systems. There is, thirdly, the interaction of these systems. There is their ultimate creator. This view does not negate the debate between the creationist and the evolutionist; it invites the two to find and nurture their proper relationships to one another and to the world in which they coexist.

"The Working Of An Organism Requires Exact Physical Laws: If it were not so, if we were organisms so sensitive that a single atom, or even a few atoms, could make a perceptual impression on our senses—Heavens, what would life be like! To stress one point: an organism of that kind would most certainly not be capable of developing the kind of orderly thought which, after passing through a long sequence of earlier stages, ultimately results in forming, among other ideas, the idea of an atom."—Erwin Schrödinger[18]

Pre-Darwinian scientific explanations along evolutionary lines have included remarkable if sometimes largely discredited views. Among these, Lamarck's theory (1809) that environmental changes can alter the characteristics of an animal, the theory of acquired characteristics, is pronounced. Lamarck's theory may be transmuted in the future to acceptability whether on an inheritable or on a single-generation, noninheritable basis.

Theories other than Lamarckianism may or may not be as fortunate. In considering the history and theories of spontaneous generation, Louis Pasteur wrote the reminder that "up to the end of the Middle Ages everyone believed in the existence of spontaneous generation."[19]

Spontaneous generation, now discredited, had at one time a strong constituent of believers. The theory of spontaneous generation was long undoubted by few scientists, Descartes and van Helmont among them, and was also accepted by some theologians, for instance the English Jesuit John Turberville Needham.* It is accepted that two different routes can share the same destination. When science and religion, or any other two kinds of focus, do not seem to share the same views by any act of translation, even rigid individuation may be a first step toward an eventual resumption of shared or compatible efforts.

While discoveries at the level of populations and genetics explain nearly all evolutionary phenomena of significance to sciencc, "whatever one calls it, that area of science concerned with evolutionary processes and phylogenetic patterns is now [1980] in a state of excitement and flux."[20] Syntheticist, transformalist, cladist, are terms to describe proponents of new branches of evolutionary thought. At the Macroevolution Conference, Chicago, 1980, a wide spectrum of researchers including geologists, paleontologists, ecologists, and population geneticists challenged the concept of "modern synthesis" which had dominated this field for four decades, Julian Huxley having coined the term in 1942. Point mutation within structural genes is a source of variability and of shifts in the gene frequencies within populations. The two essential statements of the theory of sudden point mutations, do not, argued conferees, fill the absence of transitional aids between species. The transition, as indicated by fossils, between old and new morphologies is not a smooth one. Something has not been accounted for, and the "missing link" refers not only to a particular and desirable fossil but also to a particular biological mechanism. Furthermore, paleontology finds that the principal feature of individual species is not change but stasis. Such evidence, or rather lack thereof, necessarily encourages lively discussion. That discussion, by way of extension into independent researches, continues. The emerging evolutionary picture, however respecified, describes periods of the stasis of species and of speciation, abrubt, rather than continuous, events. It accomodates both constancy and change and is

* Theologians could subscribe to this view: "for Genesis tells us, not that God created plants and most animals, directly, but that He bade the earth and waters to bring them forth: since this directive was never rescinded, there is nothing heretical in believing that the process has continued" (George Wald, "The Origin of Life," *Scientific American,* August 1954).

referred to as *punctuated equilibrium*. The point mutations of this theory allow social Darwinism to be undercut. The human process is no longer a relentless competition for survival. "Moreover, we can no longer view ourselves as the perfect result of a process of testing and refinement going back 3 million years to *Australopithecus*," (Jerry Adler with John Carey, "Enigmas of Evolution," *Newsweek*, March 29, 1982). For decades population geneticists worked to explain evolution in terms of a gradual spread of genetic change. The relationship between evolutionary and genetic studies will be both strengthening its bond and redefining its approach in the future.

The emminent scientist C. H. Waddington has written that evolutionary theory is inadequate if it states that the essential characteristics defining life are genetic transmission and natural selection. But this implies that the discovery of the gene has already been made and, for our purposes, it has not—not yet.

AGE OF CULTIVATION

During the Age of Cultivation the foundations of the "new biology" take on firmer shape. Already the base of the trunk of the tree has been evidenced. If the trunk is extended to its full height, not only the present genetical history discussed but something of the future is also visible (Table 1–2.)

Hybridization experiments begun by Rimpau and Tedin in 1890 provided the basis for results that could readily have been interpreted on Mendel's principles. The plant breeders involved, however, were ignorant of Mendel's paper. A peculiar silence, one of the strangest in the history of biology, loomed over the 34 years between Mendel's first paper and the turn of the century. The paper, published in 1866, was cited only four times during this period. One of these was in the bibliography of an article, "Crossbreeding and Hybridization," by L. H. Bailey. Bailey had never even seen the paper. This bibliographic mention of Mendel's paper served little function until years after the publication of Bailey's work, when it played a suggestive role in the rediscovery of Mendel.

The external structure and function of life still largely predominated scientific study. Heredity was as yet mysterious and the cell was barely defined when *Origin of Species* was first published. Scientists who were

Table 1-2. The Essential Tree of Genetics History.

	TRUNK OF THE TREE[a]
Future ↑	Berg (first genetic engineer)
	Kornberg (discovered path to biosynthesis)
	Watson and Crick (elucidated a structure for DNA)
	Avery (stated nucleic acid carries genetic specificity)
	Morgan (gene theory)
	Bateson (formalized the study he named genetics)
	Mendel (his rediscovery: heredity in units)
	×
	Weismann (germ plasm theory)
	Miescher (discovered nuclein-DNA)
	×
	Schleiden and Schwann (cell theory)
	Kolreuter (hybridization)
	Leeuwenhoek (first bacteriologist)
Past	Camerarius (plants are heterosexual)

[a] Tree, with two exceptions (Leeuwenhoek, bacteria; Camerarius, plants) is specific to cellular discoveries and theories. The two exceptions, however, are equally essential to genetic research and its history.

later to become recognized as geneticists were known instead in other areas of study, such as the plant physiology from which Bateson and Davenport emerged to attract others to early genetics.

The general scientific climate was rather atheistic, and rational philosophy was beginning to take shape. Science had clearly divided from religion in its search for itself, and materialistic viewpoints took over. Research during the "silence" within this period had certainly not come to a halt but was balanced on the threshold of the future.

Three years after Mendel's paper and into the long silence, a curious cellular substance was discovered. The substance was DNA. Peering into the pool of the cell the German scientist Friedrich Miescher had a virtually narcissistic panorama before him, seeing, essentially, himself and every self. The reflection in 1872 was clouded, did not even know its name, and was a long time coming to be referred to as Midas' Gold. Miescher termed the new substance nuclein. Its discovery was accidental and was not reported until two years later. Miescher did not recognize the importance of his discovery. He considered nuclein to be "nothing more than a storehouse of phosphorus for the cell" (Portugal and Cohen). At the time of Miescher's discovery protein was strongly considered to be the agent for heredity. Furthermore, it was believed that genes did not manufacture protein, rather it was protein which brought the genes to life. Not until 1889, between the appearance of germ-plasm theory and chromosome theory, was nucleic acid separated literally and figuratively from protein. Out of its parts, the whole picture of cellular activity began to achieve an essential clarity. Separation, the crucial act of molecular biology, in which the function of parts is understood in terms of how the remaining whole responds to loss of them, performed a vital function when nucleic acid was separated from protein.

"In molecular terms, the gene, which had been conceived as an active *unitary particle,* now was viewed as a segment of a large number, possibly hundreds, of base pairs in a chain of thousands of nucleotides distributed in linear order in a chromosome or comparable structure."—L. C. Dunn

With the coming of the germ-plasm and chromosome theories recognition of heredity as the transmission of a particular chemical compound prepared the way for an explanation of nuclein, or nucleic acid, as DNA, DNA in terms of genes, and the gene as the particular mechanism by which heredity is indeed transmitted.

The important theoretical role of germ plasm involved three main figures: Weismann, the theory's creator; Galton, who stimulated him; and Naegeli, who preceded him. Galton assumed "that the whole collection of units representing cells of different kinds constituted a hereditary material known as the 'stirp'* which was distinct from the body cells. The stirp of the child was thought to derive directly from the stirp of each parent, but the bodily characters of the child express only a portion of his own stirp, i.e., those units which become active, while other units remain latent in the stirp but are still transmitted to the next generation" (Dunn).

Galton's stirp theory was the direct ancestor of germ-plasm theory. Sex and body cells were rightly perceived as independent, the former functioning chiefly in inheritance as producers of germ cells. The transmitters were the stirps. It was concluded that "hereditary continuity is thus from stirp to stirp, and not from body to body, and this was a new idea. . . ." (Dunn). While stirps produced germ cells, germ cells carried idioplasm, which Naegeli (1884) stated as containing complexes of molecules. The molecule itself had just recently been conceptualized. The molecular complexes contained in Naegeli's idioplasm were viewed as determining the differentiation of cells, tissues, and organs. This is remarkable. Heredity was finally understood in terms of units. Yet, Naegeli never sought a precise location for his idioplasm nor did he try to identify it with any morphological constituent.

Weismann, meanwhile, was not the first to identify idioplasm with nuclein. In the very same year as Naegeli described idioplasm, Hertwig perceived the possible connection. Nor was Weismann alone in following Hertwig: A. von Kolliker and Eduard Strasburger also arrived at the same view. But it was August Weismann who brought the nuclear theory of heredity, germ-plasm theory, to prominence. Weismann accomplished this at a time when the behavior of nuclear material had not yet even been elucidated. His own daring helped lead into one of the most daring demonstrations of scientific prowess, genetic engineering.

Before 1895 nucleic acid was a specificity as yet virtually lost in a theoretical shuffle. Science strained to comprehend its own observations, observations defining and redefining the mechanisms of life. Now, with the Lamarckian theory of acquired characteristics suffering a rejection (it actually survived in the background long after Mendel's rediscovery, then was further submerged until finally reemerging in the 1980s), Weismann's germ-plasm theory was initiating the path to chro-

* Galton coined this term from the Latin for "root," *stirpes*.

mosome theory. E. B. Wilson stated chromosome theory in 1895. At the time his statement could not help but be prophetic:

"Now chromatin is known to be closely similar to, if not identical with a substance known as nuclein . . . which analysis shows to be a tolerably definite chemical composed of nucleic acid (a complex organic acid rich in phosphorus) and albumin. And thus we reach the remarkable conclusion that inheritance may, perhaps, be effected by the physical transmission of a particular chemical compound from parent to offspring."

Without realizing it, Wilson was talking about genes almost as they are known today. Biologists were as yet unaware of the importance of chromosomes. Wilson was, essentially, speculating. The idea of genetic particles was generally separate, in the minds of biologists, from whatever significance chromosomes might have; genetic particles were completely hypothetical. Weismann, in identifying chromosomes with genetic particles, brought their importance into the consciousness of biologists. This, Olby states, was the very point neither Naegeli nor Darwin had been prepared to accept. Therefore, it was "Weismann's innovation [which] prepared biologists for the reception of Mendelian heredity. . . ." (Olby).

"He [Weismann] attached great importance to the reduction division and he realised that here was a source of variation, for if one set of genetic elements (he called them 'ancestral germplasms') were expelled from one egg during meiosis and another set from another egg 'it follows that no two eggs can be exactly alike as regards their contained hereditary tendencies.' Hence 'the germ cells of any individual do not contain the same hereditary tendencies, but are all different, in that no two of them contain exactly the same combinations of hereditary tendencies. On this fact the well-known differences between the children of the same parent depend.'"

Occasionally in history a method of thought seems mere happenstance as long as the answer it arrives at is accurate. The thread-shaped chromosome, consisting largely of DNA and protein, lines the nucleus of every animal and plant cell. It had finally become visible as microscopes improved; still it was ordinarily visible only during mitosis or meiosis when it contracts into a thick rod. Chromosomes received their name in 1888 from Waldeyer, but as a theory did not know success until 1912.

The chromosome theory of heredity made its official appearance as "Chromosomes and Heredity" in 1910 when Thomas Hunt Morgan gave his review under that title. Chromosomes, it was held, consist of genes in linear arrangement, and by the methods of experimental breeding, the order of these genes can be mapped. This view had been developed by Morgan, E. B. Wilson, and others. Prior to this, chromosome theory had made appearances in the deductive speculations of Weismann and in a theory put forward by Theodor Boveri. Boveri based his theory on "chromosome individuality." He took this individuality to such extremes, however, that it denied, unfortunately, the "direct genetic connection" (Wilson) that exists between the chromosomes of a given nucleus.

A molecular hierarchy developed during this period, although no one yet comprehended what might ultimately be done with these molecules, or fully appreciated the extent of their importance in and among biological systems. Table 1-3 depicts a chart of the molecular hierarchy as it was then understood.

In 1902 Theodor Boveri and Carl Correns each independently noted a connection between chromosomes and plant hybridization experiments. In 1904 Boveri discovered that chromosomes are individual, highly differentiated structures. The year between 1902 and 1904 found Walter S. Sutton also discussing the role of chromosomes,

Table 1-3. The Molecular Hierarchy of Material Bodies.

ID[a]
Containing the whole architecture of germplasm.
DETERMINANT
Group of biophores; determinants. "Pieces of inheritance determine cell form, cell grouping, and, therefore, inherited characters.
BIOPHORE
Roughly equivalent to what de Vries called a pangene.

[a]Ids constituted an *Idant,* which represented the linear sequencing of the ids. The "idant" was the term Weismann used for chromosome.

The chromosome in this view was visible as preeminently functional in transmission. This new perspective led Weismann to assume the existence of reduction division.

including chromosomal pairing. Sutton made an additional statement. This statement expressed a quite new and original idea and formed the conclusion of his 1902 paper:

"I may finally call attention to the probability that the association of paternal and maternal chromosomes in pairs and their subsequent separation during the reducing division . . . may constitute the physical basis of the Mendelian law of heredity. To this subject I hope soon to return. . . ."

Chromosome theory as it had been defined by Sutton was unique and stimulating. The support of an experimentally reached chromosomal interpretation was given by Morgan to a theory put forward seven years earlier, in 1903, when Sutton did indeed return to the subject of chromosomal pairing and a physical basis for the Mendelian law of heredity. Sutton proposed that the distribution of chromosomal pairs was parallel to the distribution of segregating pairs of gene alleles at meiosis. While certain quirks existed in the whole of his theory, such as there being no allowance for crossing over between allelomorphs in the same chromosome and no appreciation of how much more there is to chromosomes than genes, "the parallel behavior of genes in inheritance and chromosomes in the nucleus made a powerful . . . case for genes being located on chromosomes" (Avers). Sutton presented his theory in the *Biological Bulletin* (April 1903) under the title "The Chromosomes in Heredity." With Sutton's paper, Dunn states, "cytogenetics . . . may be said to have begun," even though the term *cytogenetics*, which refers to a union of genetics, with its attention to the gene within the cell, and cytology, with its concern for the cell as a body, was invented much later.

Sutton's chromosome theory was assisted by Correns and by de Vries. Correns independently gave the corroborating assumption of a serial order for alleles, one set per hybridized homologue:

$$\frac{\text{a b c d e}}{\text{A B C D E}}$$

This ordering would allow for a kind of rotational recombination. Correns, however, was not referring to the exchange of blocks of alleles as stressed by Morgan and later considered in connection with crossing over. Nor did he refer to consistent loci.

Reflecting on Sutton's paper, de Vries suggested crossing over. He assumed that at synapsis an exchange takes place between the maternal and paternal. He implied that in fact units must be present in the same serial order. A year later, in 1904, Boveri gave an experimental conclusion of functional individuality of chromosomes. Chromosomes were essential to the control of development. Science was already trembling in the reverberations of a startling incentive to biology: the rediscovery of Mendel.

The overall affect of the rediscovery of Mendel's laws is well known. Specifically, the significance, if not the accuracy, of Mendelian theory is such that its rediscovery in the year 1900 marks the official birth of genetics. The circumstances that led three individuals to make this rediscovery are less well known. Tschermak, for instance, had begun research into experimental breeding (1898) in order to try to test some Darwinian ideas about the effects of self-fertilization and cross-fertilization.

In the last half of the nineteenth century there arose two enigmatic approaches to the problem of heredity: One viewed heredity through the microscopic world of the gene, the other through a still macroscopic theory of evolution. The world would seem to have been quite impatient for a reevaluation of beliefs on immediate (generational) origin and its peculiarities or characteristics. This fact formulates only part of the landscape which immediately preceded the dawn of the twentieth century, for the world had seen in the 1700s and 1800s man helpless in the face of Irresistible Grace, and then helpless in the hold of the irresistible mechanism of nature. "The mechanism of God and the mechanism of matter," writes Whitehead, "were the monstrous issues of limited metaphysics and clear logical intellect. . . . Mankind soon lost interest in Irresistible Grace; but it quickly appreciated the competent engineering which was due science." The alternations of our concentration are but a tribute to two great dilemmas intellect has in rationalizing mind and matter.

The world preparing for Mendel's rediscovery was a world desirous of extending the sheerly rational materialistic approach. Thought had received a shock from Darwin and wanted further suggestions in order to contemplate human physiological status more fully and nearly.

In 1877 Hugo de Vries found that the weather conditions of Amsterdam were unfavorable to harvesting experimental hybrids. He abandoned his work on the crossing of varieties of maize. He turned his attention to the osmotic pressure in plant cells, making valuable contributions to the plasmolytic and physiological lines along which this study led him. Only as the century began to turn did de Vries return to genetics. In 1889 his *Intracellular Pangenesis* was written, and three years later he was not only hybridizing but at last receiving desirable results from these experiments. He obtained, in 1896, a close approximation of the 3:1 ratio and demonstrated the shared heterozygotic and homozygotic constitution of dominant f_2 segregates. But this was only a beginning. Hugo de Vries was pioneering his way into the as yet unnamed and relatively little explored new world of genetics. The way was as rugged as if no trails had been laid. After 11 years of experimenting in hybridization, confident and excited with his results, de Vries wrote two papers, French and German reports, to describe his work. The French paper was already posted and the German one completed when de Vries learned that his work had been anticipated. He had had the pioneering spirit, wisdom, tolerance, and tools, and had seen the summit of his adventures only to find that the trail had in fact already been trod and had simply become overshadowed for lack of sufficient use. In a way, time had been too early for Mendel and too late for de Vries.

Hugo de Vries' discovery of the Mendelian laws of heredity was aided by the innocent, perhaps, good will of his friend in Delft, Professor Martius Whilhelm Beijerinck, who wrote to him to say, "I know that you are studying hybrids, so perhaps the enclosed reprint of the year 1865 by a certain Mendel which I happen to possess is still of some interest to you." Perhaps de Vries had previously mentioned to Beijerinck, who had also had suspicions of a hereditary operation something like unitary mutation, his notice of bibliographical reference to Mendel in Bailey's volume on plant breeding. As much as reconstruction of the matter attempts to complete the puzzle, it is impossible without further and conclusive evidence to be certain of de Vries' motives in regard to the extent of recognition he pays Mendel. How we are to interpret his words, oral and written, in regard to the matter is up for conjecture. Without Mendel's paper the law of segregation would have been known as de Vries' law. Wounded pride may indeed have been the reason for de Vries' refusal to add his signature to the Mendel Memorial Appeal of 1906. The fact of his at least publishing the results of his own experiments, for which he had waited during seven years of the repeated work that would ensure his confidence, and the following paragraph he appended to his German report were essential in the initiation of what was to become known as the study of genetics.

"From these and many other experiments I conclude that the law of segregation of hybrids in the plant kingdom, which Mendel established for peas, has a very general application and a fundamental significance for the study of the units out of which the specific characters are compounded."

What had been a silence for Mendel's paper had not been a silence in general, and the result of discoveries made between 1865 and the turning of the century led Carl Correns to state that "the intellectual labour of finding out the laws anew for oneself was so lightened that it stands far behind the work of Mendel." Correns and his contemporary, Tschermak, were anticipated not only by Mendel but also by their peer, de Vries.

Three years after de Vries approximated the 3:1 ratio, Correns and Tschermak reached the same point in their own work. Correns and Tschermak, it should be clarified, were not working together but, by coincidence, similarly and simultaneously, so much so that their work is referred to collectively. The urgency to publish their results did not occur immediately upon realization of Mendel, whose work had been carried out long ago and was in a sense, then, somewhat noncompetitive; instead, this urgency was in response to the appearance of de Vries' first report, which had been to the Paris Academy. Within two months, between March and May, the first experimentally established laws of heredity burst out of obscurity, three decades' neglect buried beneath a monument of appreciation. The laws had been confirmed not once but three times, not through collaboration but through independent research.

"The mysterious origin of cells consists in the fact that the plant develops each one that she employs from others previously existing."—Gregor Mendel

Mendel's goal was a sophisticated and predictive one. He wanted to follow the development of hybrids' progeny "in such a way as to make it possible to determine the number of different forms under which the offspring of hybrids appear—to arrange these forms with certainty according to their statistical relations'' (Mendel).

Mendel's Fundamental Concepts

1. Units—Unit characters capable of hereditary transmission constitute the mosaic known as an organism.
2. Dominance—One unit character may mask, make recessive, another when the developmental potential for both is simultaneous in the same individual.
3. Segregation—Units may segregate during reproduction, otherwise stated as homologous chromosomes separating during meiosis.

The first concept relates to Galton's ''ancestral inheritance.'' An expanded knowledge of ''units'' allows the concept to be described as follows: Genes located in the chromosomes of body cells contain hereditary potential and determine the expression of specific unit characters. The second concept, dominance, relates to Galton's ''filial regression'' and is recognized in the later concept of mosaic inheritance in which unit characters may avert the consequences of dominance by the development of the two units in different parts of the body. Segregation is associated with linkage and crossing over, and with translocation.

Mendel referred to his discovery, which was based on his study of seven pairs of unit characters evidenced in garden peas, as the law governing Pisum. Mendel rightly believed that the basis of his discovery would find applicability beyond the Pisum and the Phascolous of his study to other plants. He stated that ''we may assume that in material points an essential difference can scarcely occur, since the units in the developmental plan of organic life is beyond question.'' This applicability has extended far beyond his vision, and beyond botany. Weismann's reflections on cytological discoveries had served to prepare cytologists and hybridists for a union of their studies in a separate study. That separate study was the science William Bateson called genetics.

"Nearly eighty years ago, my father, William Bateson, became fascinated by the phenomena of symmetry and metameric regularity as exhibited in the morphology of animals and plants. It is difficult today to define precisely what he was after, but, broadly, it is clear that he believed that an entirely new concept of the nature of living things would develop from the study of such phenomena. He held, no doubt, correctly, that natural selection could not be the only determinant of the direction of evolutionary change and that the genesis of variation could not be a random matter. He therefore set out to demonstrate regularity and 'lawfulness' among the phenomena of variability."[21]

The resulting study by the English zoologist, William Bateson, was compiled and published (1894) as *Materials for the Study of Variation*. The book's subtitle, "treated with especial regard to DISCONTINUITY* in the origin of species," was an open challenge which incited the biometrician-Mendelian debate.

"Heredity and variation, the concepts of continuity between generations and of the differences which appear within that continuity, have always to be considered together. Once it was the fixity of species, the maintenance of type, and the occasional departures from type which require their association. Later, when the particulate nature of heredity became known, it was realized that only those particles or genes could be recognized which had varied or mutated to produce alternative forms. What we know about heredity, we know because of the existence of hereditary variation."[22]

It was left to another researcher, Johannsen, to reconcile the fact of continuous variation with the idea of discontinuous variation by mutation. Meanwhile, Bateson's challenge of discontinuity was accepted by his friend, W. F. R. Weldon, an Oxford zoologist. After 1900 the differences between Bateson and Weldon made them opposing leaders in the biometrician-Mendelian debate. For this to occur Mendel had first to come to the attention of the English-speaking world, and this was achieved by Bateson himself.

On May 8, 1900, Bateson took a train to London where he would deliver a paper, "Problems of Heredity as a Subject for Horticultural Investigation," to the Royal Horticultural Society, but en route to the society something happened that was to deeply affect the topic of his

* Capital letters are Bateson's.

paper. On his way he read, for the first time, Mendel's paper of 1866. It suggested what were already his own convictions and, buoyed, he incorporated it into his lecture. "From that day on," as Dunn writes, "Bateson devoted all his enthusiasm and literary gifts, which were considerable, to promulgating what he soon came to call 'Mendelism.'" He had Mendel's paper of 1866 translated into English and published in the *Journal of the Royal Horticultural Society*. The year was 1900.

"In all things," Paracelsus has written, "four elements are comingled one with the other, but in each thing one of the four is perfect and fixed." Paracelsus is referring to elements in the physical world of study, but perhaps the statement is not unfittingly seen another way. Bateson was the fourth and last of Mendel's rediscoverers. He was the vital fourth party or element in the rediscovery. Bateson fixed Mendel in the mind of science and the world (see Table 1-4). Then on July 31, 1906, in his inaugural address to the Third Conference on Hybridization and Plant Breeding convened by the Royal Horticultural Society, Bateson gave a name to the new studies in heredity, "whether of animals or of plants," and the meeting became the first conference on genetics.

"I suggest for the consideration of this congress the term Genetics, which sufficiently indicates that our labours are devoted to the elucidation of the phenomena of heredity and variation. . . ."

The theory of the gene, upon which the study of genetics focuses is understood as an early description of living units as vehicles transmit-

Table 1-4. Rediscovery of Mendel.

Bateson, 1900 Rediscovery for English-speaking World		*de Vries,* 1890s, published 1900 Initial Rediscovery
Correns	and	*Tschermak*
	Galton[a] 1875	

[a]Galton is said to have "cleared the way for acceptance" of Mendelism although he did not actually adopt Mendelian views. For assistance in understanding Galton's situation, I suggest Robert de Marrias' "The Double-Edged Effect of Sir Francis Galton" (*Journal of the History of Biology,* vol. 7, no. 1, Spring 1974).

ting heredity. It was accepted in spite of its failure to account for development and the paradox contained in its statement. The idea of unitary activity began with the "physiological units" by which Herbert Spencer explained his view of protoplasmic organization (1864). The succeeding conceptions included Darwin's "gemmules," Naegeli's "idioplasm," and Weismann's "idant" of biophores and ids, and finally Johannsen discovered in this succession of thought that the gene had been born in men's minds.* Ideas about living units continued to grow. The twentieth-century perspective differs from that of the preceding one. We continue to think in terms of these units and retain the name of gene, but we apply the word to something that would not then have been defined as it is now. In a sense, we are talking about intellectual linguistics, the evolution of a perspective.

Gene theory was more than three-quarters of a century beyond "physiological units" and "gemmules" before it gave a satisfactory explanation of reproduction, differentiation, and development. Morgan had described the theory of the gene as stating "nothing with respect to the way in which the genes are connected with the end product or character" (Dunn). The theory described genes only as vehicles for transmission. Or did it? The answer is no, but if it had, its acceptance would have perhaps avoided any embarassment. However, "the arguments for the restricted theory were so persuasive that they were accepted by most biologists, in spite of the paradox that the mechanism proposed assumed the same variety of units in all cells although the cells themselves became different" (Dunn).

Thomas Hunt Morgan evolved his concept of the gene out of chromosome theory; his concepts were largely based on his work with the fly, *Drosophila melanogaster*. The great age of experimentation with plants had ended. The great advantages of *Drosophila* were rapidity and ease of cultivation. Within two weeks several hundred offspring might be produced by a single pair. These could be readily observed during any stage of development.

The first analysis of a mutant *Drosophila* was reported by Morgan in 1910, before his "fly room" had yet been established. The results of this analysis showed a sharp contrast with expectations natural to the

* Johannsen coined the term *gene* in 1909.

applicability of Mendel's second law. Other similar results compounded the problem. Morgan therefore concluded:

"We may make a general statement and hypothesis that covers cases like these, and in fact all cases where linkage occurs: viz. that when factors lie in different chromosomes, they assort freely and give the Mendelian expectation; but when factors lie in the same chromosome they may be said to be linked and they give departures from the Mendelian ratios. The extent to which they depart from expectation will vary with different factors. I have suggested that the departures may be interpreted as the distance between the factors in question."

In the winter of 1910–11 Morgan gained the assistance of two Columbia College undergraduates, Alfred H. Sturtevant and Calvin Bridges. The "fly room" was begun. The flies were bred in milk bottles containing nutritious bananas. Larvae could pupate on support structures consisting of used envelopes from Morgan's correspondence. For classification the flies were scrutinized with a hand lens. Simple homemade installations had ingeniously been made to serve a major scientific advance.

The Belgian cytologist Janssens postulated that genes in close proximity to one another would "assort" more regularly than genes located further apart along a chromosome. In the provocative language of James Watson, "Morgan and his students were quick to exploit the implications of Janssens' still unproved theory. . . ." The theory suggested a mapping technique for the location of the relative positions of genes along chromosomes. L. C. Dunn gives the story a balanced approach:

"The basic theory of linkage as due to location of the *linked* genes in the same chromosome, and breaks in the linkage as due to *crossing over* according to the chiasmatype theory of Janssens, appeared in September, 1911, although Morgan . . . had first stated in a public lecture on July 7, 1911, at Woods Hole, Massachusetts, that: 'during segregation certain factors are more likely to remain together than to separate, not because of any attraction between them, but because they lie near together in the chromosome. . . .'"

Within four years after the "fly work" had begun 85 mutant genes in the *Drosophila* had been assigned by Morgan's group (see Table 1–5).

Table 1-5. Gene Mapping.

The following describes a portion of the genetic map of *Drosophila melanogaster,* chromosome 2.

Normal		Mutant
Long aristae	0—al	Aristaeless (short aristae)
Long wings	13—dp	Dumpy wings
Long legs	31—d	Dachs (short legs)
Gray body	48.5—b	Black body
Red eyes	54.5—pr	Purple eyes
Long wings	67—vg	Vestigial wings
Straight wings	75.5—c	Curved wings
Straight wings	99.2—a	Arc bent wings
Red eyes	104—bw	Brown eyes

Many different genes influence a single character. The following list, a portion of the 85 mutant genes reported to be in *Drosophila melanogaster* (1915), very much reflects this fact.

Group One:

Name	Region Affected	Name	Region Affected
Abnormal	Abdomen	Lethal, 13	Body, death
Bar	Eye	Miniature	Wing
Bifid	Venation	Notch	Venation
Bow	Wing	Reduplicated	Eye color
Cherry	Eye color	Ruby	Leg
Chrome	Body color	Rudimentary	Wing
Cleft	Venation	Sable	Body color
Club	Wing	Shifted	Venation
Depressed	Wing	Short	Wing
Dotted	Thorax	Skee	Wing
Eosin	Eye color	Spoon	Wing
Facet	Ommatidia	Spot	Body color
Forked	Spine	Tan	Antenna
Furrowed	Eye	Truncate	Wing
Fused	Venation	Vermilion	Eye color
Green	Body color	White	Eye color
Jaunty	Wing	Yellow	Body color
Lemon	Body color		

This was very revealing work. It could be seen that any one character such as eye color or wing shape is developed not by a one-to-one correspondence between genes but instead by a series of events, each of the series being controlled by a gene.

The developmental stages of *Drosophila* include that of the egg, three larval instars, papa, and "imagines," or adults. Homunculus had assumed the wings, not of "radiance," or the dove, but rather of a tiny fly. The wings of the fly are well remembered for having carried to new clarity the weight of the material approach to the problem of generation and maintenance.

Among Morgan's virtues were enthusiasm and an ability to work well with others. He also had a skeptical and critical attitude which greatly assisted his work. He began to doubt chromosome theory almost upon its conception. Morgan sensed, without certainty, a flow in this theory, a flow that only the future could reveal. He was concerned with the disproportionate quantities of characters to chromosomes. Since there were many more characters than chromosomes, the characters must form groups. But this did not seem to be the case. There was still another criticism to be made: "If Mendelian characters are due to the presence or absence of a specific chromosome, as Sutton's hypothesis assumes, how can we account for the fact that the tissues and organs of an animal differ from each other when all contain the same chromosome complex?" (Morgan, "Chromosomes and Heredity," 1910, Amer. Nat., 44:449–496). The paradox of similar chromosomal complexes among differing animals has continued into modern times.

Bateson had written (1902) concerning the essence of the Mendelism behind all work proceeding Mendel's rediscovery that:

> "The essential part of the discovery is the evidence that the germ cells or gametes produced by crossbred organisms in respect of given characters may be of the pure parental type and consequently incapable of transmitting the opposite character: that when such pure similar gametes of opposite sexes are united together in fertilization, the individuals so formed and their posterity are free from all taint of the cross; that there may be, in short, perfect or almost perfect discontinuity between these germs in respect of one of each pair of opposite characters."

The heredity to which the new field of genetics devoted itself came under fire when the study divided its methodology into that of the biometrician and that of the Mendelian. Sir Francis Galton's magnificent work, *Natural Inheritance* was inspiration to not one but both parties. Mendelians enjoyed his stress on discontinuous variation, while biometricians praised his introduction of statistics to biological studies.

The concept of heredity was itself largely due to Galton for, in introducing the concept of nature versus nurture, he provided a definition for the word *heredity* which operated on a much richer level than the definitions preceding it. Even in the 1980s nature and nurture, along with the previously stated problem about chromosomes, has remained intriguing and as yet unexplained in any truly clarifying manner. But major mysteries had been unveiled, the veils yet falling away.

AGE OF PHYSIOLOGICAL STUDY

"As you no doubt know, nucleic acid is an extremely important constituent of cells and it seems to us that it would be very valuable if this could be followed up in detail."—Sir John Randall's letter to Rosalind Franklin, 1950*

The issue of nature versus nurture was given a new dimension in the 1900s. Heredity was no longer a question; it was an answer. Nucleic acid became more than a vaguely perplexing coincidence of research. The lines of heredity took shape. Their geometry produced a helical figure. A study of the spiral became a study of life, not just the qualities evident throughout life, but the physical properties inherent to life itself.

Given practical laboratory experience and intuitive insight James D. Watson dreamt a dream. He awoke with the first acceptable formula for the structure and the function of DNA. And with this the world began to wake to a collective focus on the double helix.

In 1944 O. T. Avery made the momentous discovery "that the hereditary properties of pneumonia bacteria can be specifically altered by the addition of carefully prepared DNA of high molecular weight." At least 12 persons later became involved in the struggle to reveal the essential characteristics of heredity in terms of what came to be known as DNA. In the mammoth volume, *The Eighth Day of Creation,* a wonderful, informative source that deals exclusively if rather chattingly with the development of genetics during the 1900s, these "Chief Persons of the Comedy" of discovery are aptly described. They include Sir Laurence Bragg, Max Perutz, and the limelighted Francis

*See Judson's *The Eigth Day of Creation,* p. 103.

Crick and James Watson. We proceed from those persons in Cambridge, England, to those in London, and the turbulent relationship between Maurice Wilkins and Rosalind Franklin. Remaining are Erwin Chargaff, New York; Salvador Luria, Indiana and Illinois; Linus Pauling, California; and Max Delbrück, a founder of molecular genetics. Independent research across the globe devoted intense concentration on the chemical nature of heredity. A competitive and almost coincidentally collaborative effort pressed the study forward.

Both the team as a whole and its individual constituents can be considered. Such a group generally has the potential to support itself as a whole or any of its individuals. Exactly how discovery will emerge is always an unknown. Pauling most certainly expected the elucidation of the structure and function of DNA to be finalized, if in a working hypothesis, by a single individual, for he suspected, and not without good reason, that individual would be himself. In fact, the most viable suggestion for a structure for DNA came instead from a subteam, the team of Watson and Crick. And this subteam would not have succeeded without the particular contributions of one of its individuals, James Watson. Judson writes of scientific individualism that "it must be said that in the process of discovery there comes a unique moment: where great confusion reigned, the shape of an answer springs out—or at least the form of the question." The insight occurs in the prepared mind of some one person. The exaltation of that moment is compared by Judson to the satori of the Zen adept. The process of discovery was now, in the early to middle 1900s, truly under way. In 1953 Watson and Crick gave science a rudimentary structure, something to work with, a figure that has undergone at least as much scrutiny and refinement as it did when it was just beginning to take shape in men's minds.

The evolving figure of the double helix has been surrounded by enthusiasm, revision, conflict, and controversy. To do justice to the fantastic extent and range of discoveries of the twentieth century, especially in the latter half of the century, would require a separate volume. For a feeling of the nature of the time I refer the reader to Appendix VI, which contains exerpts from important papers and from the correspondence of, among others, Francis Crick.

"The most incomprehensible thing about the world is that it is comprehensible."—Albert Einstein

The study of the double helix of DNA has made heredity comprehensible. The double helix is the awkward caduceus. It is not the spiral of spiritual life, but its counterpart, the helical spiral of physical life. While my goal has been to present facts more than fervor, clearly the latter may be imagined. "The elucidation of the structure of deoxzyribonucleicacid," writes Judson, "had been a concentration of forces, a siege, a conquest."

Watson and Crick were studying not just one large molecule, but one large molecule composed of many smaller molecules. The work of

Table 1-6. The Amino Acids from Which Natural Proteins Are Synthesized.[a]

Gamow's Most Abundant Twenty[b]	Crick's and Watson's "Magic Twenty"[c]	Conventional Abbreviation
Glycine (6)	Glycine	Gly
Alanine (10)	Alanine	Ala
Valine (7)	Valine	Val
Leucine (2)	Leucine	Leu
Isoleucine (12)	Isoleucine	Ile
Serine (4)	Serine	Ser
Threonine (11)	Threonine	Thr
Aspartic acid (3)	Aspartic acid	Asp
	Asparagine	Asn
Glutamic acid (1)	Glutamic acid	Glu
	Glutamine	Gln
Lysine (5)	Lysine	Lys
Arginine (9)	Arginine	Arg
	Cysteine	Cys
Cystine (15)	—	—
Cysteic acid (19)	—	—
Methionine (17)	Methionine	Met
Phenylalanine (14)	Phenylalanine	Phe
Tyrosine (13)	Tyrosine	Tyr
Tryptophane (18)	Tryptophan	Trp
Histidine (16)	Histidine	His
Proline (8)	Proline	Pro
Hydroxyproline	—	—

[a] Source for table and concluding footnotes: Judson, Horace Freedland, *The Eighth Day of Creation,* Simon and Schuster, 1980.

[b] Gamow's remaining five, less frequent: norvaline (21), hydroxyglutamic acid (22), asparagine (23), glutamine (24), cannine (25).

[c] Crick's and Watson's rejected candidates: cystine, hydroxyproline, hydroxyllsine, phosphoserine, diaminopimelic acid, thyroxine and related molecules (that is, the iodinated and brominated tyrosines).

Watson and Crick was of that complex nature familiar to navigators of unchartered territory and included a measure of guesswork the results of which were not always immediately recognized as correct although their value was often shown in time, as in the standard set of amino acids arrived at by the team (see Table 1-6). Studies in genetics progressed rapidly from the 1953 publication of the Watson-Crick findings on the structure of DNA. In 1956 the path to the biosynthesis of DNA was discovered, just three years later. In 1967 Kornberg, Goulian, and Sinsheimer performed the first DNA synthesis. We are given a complex of individuals who, together, express a single heroic endeavor, as if they were one hero. We perceive our hero as:

> The man who has penetrated into an obscure region of research, who has attempted to examine stone by stone in their production and application, who has given us both ground plans and elevations of some plant and other more complicated structures in which each element is marked with the number its architect intended for it.*

This archetypal man has given us the architecture of biological structures in details intimate for nearly all forms of life. He has explored the maintanence of these structural details and has ventured their manipulation. He is one who continues his examination until his knowledge is so great that we assume from him promises for a new future.

The characteristics of the cell upon which this figure's study is focused are the characteristics which have shaped biological life for countless centuries.

*Mendel of C. Naegeli.

2
Characteristics of the Cell

"... In our bodies is repeated the structure of the Whole."[1]

Pythagoras, in considering the formation of the three-dimensional world of solid objects, observed that the Three creates the plane, the fourth aspect completing the solid by adding to it the third dimension. As vessels of a fluid world, being neither flat images drawn across a canvas nor ethereal beings blown on a breeze between sky and earth, we ourselves appear to be three-dimensional. With simple regard of the numbers three and four when considering genetics, the human body, the living body, indeed the biological life form becomes the greatest of expansions of Platonic solids. The genetic code constructing these organisms is read in groups of three; each gene is composed of three couples, whose individual components are selected from four possible types. As with the aspect of four concerning the dimensions of the underground bases of the major Egyptian pyramids, the fact of four is less visible than the resulting three, whether it be the three couples of the gene or the ratio between the visible height of the triangular sides of the pyramid and its submerged, invisible base.

The symbolic dimensions of genetic arithmetic have been explored by the English crystallographer William Thomas Astbury, and also by the analytical psychologist Marie Louise von-Franz, who writes that "the I Ching oracle, whose rudiments originated in the 'revelations' of lowly vertebrates, symbolizing the unconscious, is based on a double constellation of triple lines. . . ." The arrangement of these lines depict the basic structure of the whole of existence. When doubled into a hexagram they signify 64 possible aspects of *unus mundus* (C. G. Jung's term for the unity of existence underlying the duality of psyche and matter) and, as von-Franz continues, "one cannot escape the impression that these numerical combinations are introspective representations of fundamental processes in our psychological nature; indeed, the modern science of genetics has discovered that the biological genetic code of DNA consists of four bases (adenine A, thymine T, guanine G, and cytosine C) combined into three. These groups constitute the code for building up the twenty amino acids that form all our bodily proteins. From the four bases sixty-four different triplets come into being."

Messenger RNA also uses triplets to form the basic figures of its code, and "in these genetic findings we are confronted with an exchange of 'information' in living cells that corresponds exactly to the structure of I Ching hexagrams. This astonishing correspondence seems, more than any other evidence, to substantiate Jung's hypothesis that number regulates both psyche and matter."[2]*

Numerology, the study of number symbolism, can at the present time hardly be said to receive uncritical respect, and has never again seen so great a day as that in which Pythagoras attributed the origin of life to number, but it is not entirely ignored by members of the established professions, such as psychologists and scientists. The crystallographer Astbury announced, if cautiously, his relationship to numerology at the first major biological conference in Europe after the second world war. "Astbury's language," writes Judson, "was rousing and certain passages stuck in the mind." These passages included the following:

"Right from the beginning of this X-ray study of the nucleic acids we have drawn attention to the fact that the spacing. . . of the nucleotides along the

* "The same numerical model, a pattern underlying the basic processes of human memory and transmission (and thereby also the substratum of our entire conscious processes), has been discovered, first in China through an introspective examination of the unconscious psyche, and then in the West through genetic research into the living cell" (von-Franz, ibid.).

nucleic acid column is to all intents and purposes the same as the distance from one side chain to the next along an extended polypeptide, and even at the risk of being accused of dabbling in numerology I should like to say again that I believe this to be no mere coincidence. *In a sense the evolution of biological molecules has been one long story of numerology,* a sifting and sorting of shapes and sizes and a progressive selection of molecular patterns that has been going on throughout the ages. Biosynthesis is supremely a question of fitting molecules or parts of molecules one against another, and one of the great biological developments of our time is the realization that probably the most fundamental interaction of all is that between the proteins and the nucleic acids. . . . The correspondence between internucleotide spacing and inter-side chain spacing . . . is not an arithmetical accident but a stereochemical correlation of deep significance."*

Astbury was the only crystallographer at the meeting. He was the one man concerned with three-dimensional molecular structures. His delivery is most unfortunately described as "the time when he regressed to the notion that DNA was built up, as meaninglessly as a starch, from multiple sets of four nucleotides."[3] Certainly Astbury's passage, as cited, is the reverse of meaningless.

Astbury suffered for his reference to numerology. Numerology is a subtle science not fully developed even after many centuries of exploration. The nature of its symbolism is not fully appreciated and it is susceptible to loose interpretation. It is a subject that receives virtually no encouragement, nor credit for refinement, but is more often ridiculed. Instead of being perceived as a clue "Astbury's numerology" was considered a distraction after it was viewed in light of the coding problem as a certain Gamow expressed it. But this was only after a quite historically unusual, if brief, entrancement with the possibilities suggested in what Astbury was saying. The enthusiasts had at one time included Francis Crick, who confronted, though not without frustration, the enchanting significance of the near identity between the distances along the polypeptide and the polynucleotide.

"I want to consider two aspects of the DNA structure. Firstly its dimensions; secondly its chemical character.

* William T. Astbury, "X-ray Studies of Nucleic Acids," in Symp. Soc. Exp. Biol. 1, held in 1946, p. 70.

the numbers 3 4

"The dimensional side is soon disposed of. In the 'paracrystalline' form of DNA (Structure B) we have one base pair every 3.4 Å in the fibre direction. A fully extended polypeptide chain measured about 3.7 Å from one amino acid to the next. Therefore it is argued that not more than one base pair can, on the average, be matched with an amino acid. If we go up the outside of the helix the position is worse, since the distance per base-pair is now greater, perhaps twice as great.

"I want to point out that this argument, though powerful, is not completely water-tight."—Francis Crick, "On Degenerate Templates and the Adaptor Hypothesis: A Note for the RNA Tie Club," 1955.

Crick wondered if DNA, acting as a template, might portray bases tilted closely together as was known to occur in the crystalline A form. This inquiry led Crick to realize the potential appropriateness of the formation of hydrogen bonds with complementary base pairs for the structure of an adaptor molecule. This would make the adaptor molecule structurally similar to DNA. Judson notes that this reference to DNA represented perhaps the most striking feature of the adaptor hypothesis, for Crick had "generalized the principle of complementary base pairing, bringing it forward from DNA to this next step." But the adaptor hypothesis complicated the coding problem. Crick began to consider that a variety of coding schemes would be structurally possible. Astbury's numerology was no longer enticing.

Analogies, promising or amusing, are provided by numerologists who might perhaps find it a curious fact that the names Watson and Crick bear out the theme of four and three respectively, the first letters of these names being located at the fourth and third positions in from either end of the alphabet. The geometry of the cell, meanwhile, provides ready access into the symbolic dimension of genetics, a side benefit of its literal importance.

Cellular geometry is of vital importance, as anyone familiar with sickle-cell anemia is well aware. The crescent shape of the sickle cell, analogous to the crescent moon waning into a period of darkness, is a devastating abnormality. It makes the full-bodied quality of the healthy, exuberant individual with normal circular or disklike cells far preferable. But there is another kind of geometry which we can percieve in the cell.

A mandalic quality in the structure of one of the largest cells of the human body, the egg, gives the viewer pause to wonder. The spiral quality of the double helix of DNA gives pause to think. In cross sections of some stones, including human gallstones, a visual likeness is borne to X-ray diffraction photographs of DNA and to certain mandalas. If it is true that "in our bodies is repeated the structure of the Whole," then this likeness is perhaps more than a superficial visual similarity and research may one day provide revealing data by a comparison of these structures.

Before it was visualized in detail in the Watson-Crick model, the double helix was seen by an X-ray diffraction photograph. Similar photographs have since described DNA in its A form and also with a left-handed helix. In the right-handed B form, as depicted in the well known early X-ray diffraction photograph by Rosalind Franklin, there is visible the traditional "X" cross. The comparatively modern cross (in the shape of a "t") is implied by the left-handed helix. The configuration of the B form of DNA, an "X", readily illustrates the outstretched human body. The core and branches of an encircled diamond, they also indicate ways to perceive intricate and symbolic forms, such as the hexagon, and become formulative to the pentagon especially when the photograph of the B form is visualized on its side. Considerations such as these, naive and incomplete compared with many others of a similar nature, are derivatives of the inescapable spiral symbolism by which the student open to the assistive device of imagination gleans the creativity that gives rise to some of the greatest discoveries made by humankind. One example is found in Kepler's employment of Platonico-Pythagorean ideas about correspondences between the orbits of planets and musical intervals to establish the famous astronomical laws of his name. In more recent times, when creative thinking has been craved in every profession, Watson's dream concerning the double helix, described in his book, *The Double Helix,* is an example of how the creative realm has assisted genetics. The perception and the employment of this mechanism will undoubtedly provide society with a variety of creative solutions to its problems in the future, for the employment of intuition and imagination is already increasing, however slowly, on a number of levels, within many professions, and in the populace at large. The realization of analogies and the ability to translate concepts between the subtle and coarse realms or between two fields of study have led to the creation of a vast quantity of the arts and the tools which characterize civilization.

Within the mandalic quality of the cell is the double helix, a form of spiral that is a highly symbolic shape having regard to the spine, the harmonized spine of duality, or the caduceus; it is the staff of life, which may be viewed as Jacob's ladder or as the cabalistic tree of life among whose branches every person is evidenced at one stage or another of a great climb. That this statement is an oversimplification does not mean that it refers to a wholly unrealistic assimilation of images either for the symbologist elaborating its points, or even for the scientist confronted with them coincidentally and indirectly. The researchers and laypersons studying genetics are studying ancient and archetypal images reflecting mystical approaches to life, and are doing so coincidentally to the manifestation of these images in the psyche of growing numbers of people. That work with ancient and archetypal images is occurring in a materialistic science speaks of the tenacious persistence of these images, their notorious refusal to remain buried.

The structure of DNA provides material for varied and symbolic interpretations. An interior sun is evidenced by the relationship of the single-stranded version of the biological helix to the shape of a spiral. The difference between a spiral (conic) and a helix (cylindrical) is readily apparent when one considers the distance between the position of the viewer and that which is viewed. Consider the vertex of that which is viewed to be the sun in the sky. Any two rays of the sun can be considered as generating the sweep of a cone. Viewed from a great distance the rays will appear, however, to be parallel. From the perspective of this distance the rays are no longer the spiralling cone they really are but a cylinder, as the rays of the sun appear when viewed from earth. The illusionary cylinder is a shape representing the winding curves of a helix. The cone, on the other hand, the actual geometrical manifestation of the rays, is recognized as an unfolding spiral. The cylindrical helix, by this line of thought, is actually a spiral, or cone. The rays of the sun appear cylindrical at a point, such as earth, that is a great distance from their point of origin; closer up their true nature, that forming a cone, becomes apparent. The helix, therefore, as an illusionary cylinder, represents a point of origin quite distant from the viewer. Symbolically, this could be accounted for as a distorted perception of the mystic spiral, manifest as materialism or the world of the physical. Another perspective allows the helix to be the essence radiating physical energy throughout the life span of an organism, so that when the helix is active, radiating, it may be considered near, unfolding the physical life, as spiraling, conical, the emanation of an active interior, physiological sun. Spiral unfoldment in leaf growth as evidenced in studies in phyllotaxis provides an example of a more literal approach to focuses on the cone.

Referring to the spiral as having "little or no direct relationship to the phenomena of growth," Matila Ghyka describes the spiral as an "attribute rather than a form." The spiral is, he states, transitory. The helix, recognized in geometry as the fourth form of the spiral, has an intimate relationship to the phenomena of biological growth. If the spiral is transitory, just as nearness or clarity may fluctuate, it speaks of change. As we know the spiral in genetical and mystical terms it necessarily refers to growth: What nurtures externally (sun) refers to what is internally in a state of becoming, both physically and psychically.

While the body may be finite it holds an infinite potential for variety, as witnessed in the variety of species, variations within species, and in the hopes of genetic engineering. When the three-dimensional spiral is completed to express the spherical vortex it embraces the double helix:[4] "The world materializes and man spiritualizes along the same spiral."

Relating the structure of the cell to a structure symbolic of consciousness, such as the spiral, may lead one to wonder about the consciousness of cells themselves. Lewis Thomas has attributed a consciousness to cells. The poet Robert Bly does not exclude this possibility from his own reflections on the cell. But Richard Dawkins has received much attention for attributing to genes a quality, selfishness, whether they have consciousness or not.

The selfishness of genes confronts the individual struggle. Writing about Darwinism in *New Scientist* (February 21, 1980), S. J. Gould stated natural selection as described from individuals struggling "for their own benefits (in modern terms, for the transmission of their genes . . .)." Dawkins devoted an entire book to the new concept of *The Selfish Gene*. He opened the door to a new perspective on the gene. Subsequently, the topic of a volume 284 issue of *Nature* announced itself as "Selfish DNA." The title was centered on a golden cover and surrounded by an illustration of a circle of strands of weaving DNA with four animal heads, including the snake or dragon, their jaws open to consume the tail of their forebearers in the cycle. In light of mythology and the archetypal quality of the structure of DNA, this illustration with its tail-eating serpent, or Uroborus as it is known as an archetypal image, further signifies the philosophical dimensions of studies in genetics. In mythology the Uroborus describes a situation in which to consume the past in the head of the new is to both accept and transform it and this, it may be suggested, is exactly what every organism does at birth in terms of its own genome and the gene pool of its species. The reason for the quality of this particular cover illustration is that the concept of the selfish gene had been picked up by various researchers, including F. H. C. Crick, and two articles reviewing the subject were contained in the issue. Biochemists W. Ford Doolittle and Carmen Sapienza of the Department of Biochemistry, Dalhousie University, Nova Scotia, wrote that:

> "The assertion that organisms are simply DNA's way of producing more DNA has been made so often that it is hard to remember who made it first. Certainly, Dawkins has provided the most forceful and uncompromising recent statement of this position, as well as of the position that it is the gene, and not the individual or the population, upon which natural selection acts."[5]

Doolittle and Sapienza stated that "natural selection operating within genomes will inevitably result in the appearance of DNAs with no phenotypic expression whose only 'function' is survival within genomes. Prokaryotic transposable elements and eukaryotic middle-repetitive sequences can be seen as such DNA's, and thus no phenotypic or evolutionary function need be assigned to them." What Doolittle and Sapienza went on to propose was the first systematic approach to explaining various classes of DNA requiring unique explanation. In the course of their research they found that:

> "Although DNA sequences which contribute to organismal phenotypic fitness or evolutionary adaptability indirectly increase their own chances of preservation, and may be maintained by classical phenotype selection, the only selection pressure which DNAs experience directly is the pressure to survive within cells. If there are ways in which mutation can increase the prob-

ability of survival within cells without effect on organismal phenotype, then sequences whose only 'function' is self-preservation will inevitably arise and be maintained by what we call 'non-phenotypic selection.' Furthermore, if it can be shown that a given gene (region of DNA) or class of genes (regions) has evolved a strategy which increases its probability of survival within cells, then no additional (phenotypic) explanation or its origin or continued existence is required.

"This proposal is not altogether new; Dawkins, Crick, and Bodmer have briefly alluded to it."[6]

The issue of *Nature* concerned with this topic contained another review of selfish DNA. This review, prepared by L. E. Orgel and F. H. C. Crick, of the Salk Institute, had as its object an attempt "to make widely known the idea of Selfish DNA" to the spectrum of scientists who had yet remained uninformed. Addressing itself to mechanisms for the spread, quantities, and locations of selfish DNA, its inheritability, its ability to acquire special functions, and the effects of such generally nonspecific DNA on organisms, the review furnished a very clear definition of this new class of DNA. Because this kind of DNA has come to receive a great deal of attention, although its full implications are as yet unclear, and because the Orgel-Crick definition is so helpful, this definition is reprinted here in full, as follows.

"The theory of natural selection, in its more general formulation, deals with the competition between replicating entities. It shows that, in such a competition, the more efficient replicators increase in number at the expense of their less efficient competitors. After a sufficient time, only the most efficient replicators survive. The idea of selfish DNA is firmly based on this general theory of natural selection in an unfamiliar context.

"The familiar neo-Darwinian theory of natural selection is concerned with the competition between organisms in a population. At the level of molecular genetics it provides an explanation of the spread of 'useful' genes or DNA sequences within a population. Organisms that carry a gene that contributes positively to fitness tend to increase their representation at the expense of organisms lacking that gene. In time, only those organisms that carry the useful gene survive. Natural selection also predicts the spread of a gene or other DNA sequence within a single genome, provided certain conditions are satisfied. If an organism carrying several copies of the sequence is fitter than an organism carrying a single copy, and if mechanisms exist for the multiplication of the relevant sequence, then natural selection must lead to the emergence of a population in which the sequence is represented several times in every genome.

"The idea of selfish DNA is different. It is again concerned with the spread of a given DNA within the genome. However, in the case of selfish DNA, the sequence which spreads makes no contribution* to the phenotype of the organism, except insofar as it is a slight burden to the cell that contains it. Selfish DNA sequences may be transcribed in some cases and not in others. The spread of selfish DNA sequences within the genome can be compared to the spread of a not-too-harmful parasite within its host."[7]

DNA in higher organisms usually falls into two classes. One of these is specific; the other, by comparison, is nonspecific. It has been suggested that nonspecific DNA originated by the spreading of sequences having little or no effect on the phenotype. This possibility is explored by Orgel and Crick in their 1980 review. That the further study of nonspecific DNA, much of which may be regarded as selfish DNA, "could well make sense of many of the puzzles and paradoxes which have arisen over the last 10 or 15 years"[8] is a valuable statement in terms of biology, and also in terms of the philosophies describing the paradoxical archetype of the double serpent harmonized in the caduceus of equilibrium and in the successful experience of the Uroborus; each of the three, DNA, caduceus, Uroborus, is an expression of the essential qualities of one form, often seen as the spine. In the first instance, DNA, selfishness is manifested perhaps ineffectually. The two latter images reflect selfishness, whether or not this quality is at times necessary and beneficial, only in the course of their downfalls or pitfalls, often resulting in ineffectuality or negative effects in medicine (caduceus) and in the spiritual/psychological quest (Uroborus). While it is not my intent here to defend the place of archetypes in culture and science, the mediated translation of research between biology and philosophy may be perceived as a viable potential for future developments in these areas, as physics became to genetics in the early part of the twentieth century. The order of one lends perspective on the order of the other.

*This has, however, been debated.

"A living cell is the epitome of ingenious design; although highly complex, it functions with orderliness and efficiency."—Wai Tiu Cheun*

The overall structure of the cell may itself be physically spiraled in terms of the myelin sheath of its membrane, as in the Schwann cell. The membrane is invisible under a light microscope. The world of the fluid cell permits an access not immediately revealing. To an extraterrestrial giant viewing the earth's oceans, the fish are barely mere shadows; to the individual mind looking in upon the self, not infrequently a pool, whether or not it is narcissistic, or a veil protects the parts of self from premature view. Often a plunge or a probe suffices to elucidate interiors. In the study of the cell the probe has revealed a remarkable world. Perusal of a possible pattern of evolution of the cell, as suggested by J. David Robertson in 1962, leads to a familiar schematic representation of the cell. Robertson describes four major steps in the evolution of the cell. In the first step, the cell appears to be simply a circular core, matrix material bounded by a unit membrane. From this core five forms emerge as peninsulas, some of the matrix material having pushed out from the center to form pseudopodia, unevenly square or rectangular in appearance, like the shapes of puzzle pieces. These are the secondary structures that will complete the core by encircling it. In the third stage, evaginated membranes fold back to make paired nuclear membranes and a primitive endoplasmic reticulum. By the fourth stage the endoplasmic reticulum exhibits much progress in its development. At this stage in the possible evolution of the cell, material that was at first completely outside the cell, peninsular, is now included in its volume; topologically it remains outside, essentially it has come within; a bridge has been constructed. In the nucleus of the cell DNA weaves its double helix. The constituents of the cell, the mitochondria, ribosomes, etc., looking like the odd, rounded shapes familiar to aquatic life, may be discerned with the help of microscopy. A glossary of cell components, their structures and functions, is given in Table 2–1.

The form of cells is the result of various forces. Surface tension, for instance, exerts a great effect on the form of cells. Surface tension is

* Cheun, Wai Yiu, "Calmodulin Plays a Pivotal Role in Cellular Regulation," *Science,* vol. 207, January 4, 1980.

Table 2-1. Glossary of Cell Components.[a]

CELL COMPONENT	STRUCTURE	FUNCTION	SIZE
DNA (deoxyribonucleic acid)	A polymer molecule in the form of a double-stranded helix containing many thousands of subunits.	Contains genetic information coded in sequences of subunits called bases.	Diameter: 20 Å (angstrom) Length: Several thousand Å up to several millimeters.
mRNA (messenger form of ribonucleic acid)	A single-stranded polymer molecule containing hundreds of subunits.	Transcribes from DNA the information needed to make a protein molecule and carries it to site of protein synthesis.	Diameter: 10–15 Å Length: 1000 to several thousand Å.
tRNA (transfer, or soluble form of ribonucleic acid)	A single-stranded polymer molecule containing approximately 70 subunits. May be folded into a double helix in some regions.	Conveys specific amino acids to site of protein synthesis. Each amino acid has its own type of tRNA.	Length: 250 Å unfolded.
Ribosome	A globular structure consisting of 40 percent protein and 60 percent RNA.	Collaborates with mRNA to link together amino acids delivered by tRNA, thereby creating proteins.	Diameter: approximately 230 Å.
Polyribosomes or Polysomes	Strings of ribosome temporarily held together by mRNA	Provide actual mechanism of protein synthesis.	Length: varies with length of mRNA. holding ribosomes together.

The production of protein molecules is the end result of the collaboration of these components.

[a] Source: Rich, Alexander, "Polyribosomes," *Scientific American,* December 1963.

due primarily to molecular force. These molecules are helical, like DNA, large (macromolecules), and they are regulatory, or they are tautomeric.

Essential to the structure and formation of cells are the class of molecules known as proteins. The primary structure of a protein is a sequence of nitrogen-containing compounds, the amino acids. This primary condition evolves into a quaternary structure of protein-protein interactions through two other structural steps. Of these two structural steps, tertiary structure, which is the folding of the chain, is naturally proceeded by helix formation.

The protein is neither a long and extended two-dimensional chain, nor is it simply a randomly coiled structure. Both order and disorder are, in this sense, present within its overall organization. Some portions of the polypeptide chain are neatly coiled (order); others are uncoiled though neatly folded (comparative disorder). As a whole, therefore, the unpurified parallels of the caduceus are expressed.

The helix structuring a protein is perceived by methods of chemical amino acid sequencing analysis and by the X-ray diffraction photography of protein crystals. The geometry of the helix turns many Platonic solids into aquatic molecular vehicles, extending the realm of their symbolic definitions. Several kinds of helices and forms of DNA are known to exist; exhibited may be the A, B, C, form of helix, supercoiled DNA, or circular DNA, for instance.

Proteins were once thought to be the hereditary mechanism. They are not DNA. They are a population of substances, however, that are coded by DNA and vital to organisms. Their individual characteristics vary, as do their ancestries. Table 2–2 describes a variety of proteins related by common ancestry.

Proteins are described by studies in organic chemistry as having molecular weights ranging from approximately 6000 to over 1,000,000. They are macromolecules. The organic chemistry by which they may be studied represents a field of interest derived from the early and now seemingly simple concept that substances of either plant or animal origin differ from substances of mineral origin (inorganic substances). In the middle 1800s this idea changed. The idea that organic substances could be synthesized only by living organisms was gradually discounted. Many natural products have been synthesized in the laboratory. Countless related materials have been made that have not been recognized to exist in nature. In all of these compounds carbon is present. More than a million carbon compounds are known. Carbon has an exceptional ability to catenate. It is because of this ability that carbon can form an unusually large number of compounds. The term organic chemistry represents a convenient division of chemistry. While the term has been retained, however, it has come to be commonly definable as the chemistry of carbon and its compounds, or hydrocarbons and their derivatives (to exclude its inorganic compounds). Carbon is a constituent of protein.

Table 2-2. Proteins Related by Common Ancestry and Their Percent Identity.[a]

Related Proteins		Percent Identity
Carboxypeptidase A (bovine)	Carboxypeptidase B (bovine)	47
Hemoglobin α-chain (human)	Hemoglobin β-chain (human)	45
Chymotrypsinogen A (bovine)	Trypsinogen (bovine)	42
Lysozyme (chicken)	Lactalbumin (bovine)	35
t Antigen (SV40)	t Antigen (polyoma)	29
Myoglobin (human)	Hemoglobin α-chain (human)	28
Azurin (*Pseudomonas aeruginosa*)	Plastocyanin (spinach)	24
α Lytic protease (Myxobacter)	Chymotrypsinogen A (bovine)	18
Hin (*Salmonella typhimurium*)	TnpR (Tn3)	36

[a] Source: Simon, Melvin, et al., "Phase Variation: Evolution of a Controlling Element," *Science*, vol. 209, no. 4463, September 19, 1980. Copyright 1980 by the American Association for the Advancement of Science.

Protcins arc formed from simple compounds, the α-amino acids. An α-amino acid is a carboxylic acid having an amino group, $—NH_2$, bonded to the C atom next to the carboxyl group, —COOH. The position of the amino group is denoted by the designation α (alpha). The C atom adjacent to the carboxyl group is the α-carbon atom. The general formula of the compounds is depicted below.

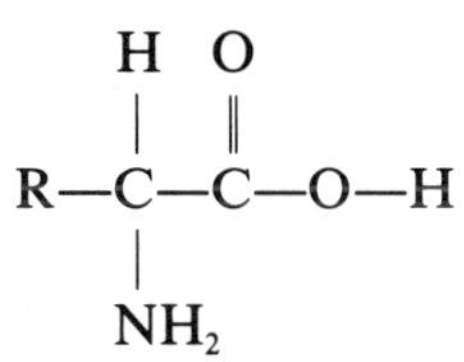

The $—NH_2$ group is basic. It will react with acids to form $—NH_3^+$. At low pH's an amino acid exists as a cation, moving toward the negative pole of an electric field. The —COOH group is acidic. It forms —COO^- with bases. At high pHs amino acids are anions moving toward the positive pole of an electric field. Internal neutralization occurs when the proton from the —COOH group moves to the —NH_2 group. Internal neutralization produces a result known by the thoroughly unneutral sounding name zwitterion.

One of the most intriguing and recent researches into protein has revealed the "split gene." In organisms higher than bacteria genetic messages coding for proteins are frequently interrupted. Discontinuous genes are prevalent in higher organisms, in which most genes are, in fact, discontinuous. The DNA that codes for protein is interrupted, for instance. It is interrupted by noncoding sequences which are far from functionless. Their transcripts are excised to make the messenger RNA that is translated into proteins. Split genes were discovered by Pierre Chambon and his colleagues in 1977.

"When my colleagues at the Laboratory of the Molecular Genetics of Eukaryotes in Strasbourg and I undertook the investigation that led to the discovery of split genes, we were not actually investigating the structure of the gene as such. We were engaged in a study of cell differentiation. In particular we hoped to learn how female sex hormones—the estrogens and the progestins—control the differentiation of cells in the oviduct of a laying hen and the expression of the gene for ovalbumin, the major egg-white protein."[9]

The researchers perceived multiple radioactive bands yielded by a single ovalbumin gene, rather than the single band typically expected to represent a genomic fragment. These multiple bands ultimately implied the possibility of split genes. The researchers could hardly recognize this at the time of their discovery. It was most frequently suggested to them that what was seen here was an artifact of blotting or hybridization.

"The changing concept of the gene is a thread that runs through the entire history of modern biology. . . . [Between the 1950s and 1970s] it became clear that in bacteria, at least, a discrete, contiguous stretch of DNA is a 'structural' gene encoding the genetic information to specify the manufacture of a single protein, and that the linear sequence of the subunits called nucleotides in this stretch of DNA corresponds directly to the linear sequence of amino acids in the protein. This principle of colinearity was assumed to apply not only in bacteria but also in higher cells. In 1977 that turned out not to be true. The organization of the gene in mammals, birds and amphibians is fundamentally different from the organization in bacteria: most of the genes are in effect split. The discovery of split genes, along with some other findings of recent years, shows that the genetic apparatus is more complex, more variable and more dynamic than any of us had suspected."[10]

The structure of genes is necessarily a subject of great interest. The gene is a sequence of nucleotides in polymer deoxyribonucleic acid. It may be altered. Among the ways it may be altered is natural mutation. In genetic engineering the alteration may be achieved by the insertion and deletion of single nucleotides. The gene is also altered by recombinations between pairs of adjacent nucleotides along the polynucleotide chain. The basic function of the gene lies in its control of the structure of polypeptide chains. Certain genes control the primary structure of individual and specific protein molecules. Mutations in a structural gene cause changes in the primary structure of the protein molecule.

A gene is a short length of the threadlike chromosome. The gene and the polypeptide chain in the principle of their one-to-one relationship are both linear structures and are colinear with each other. Many different genes constitute DNA. Each gene is replicated exactly and repeatedly during growth. The different forms that a given gene can change into by mutation constitute a series of alleles, although interactions between nonallelic genes also exist.

The interactions between genes, as well as within a gene have not always been clearly understood. "For many years," writes Watson, "it was generally thought that crossing over occurred between genes, not within the genes themselves."[11] The chromosome was viewed as a linear collection of genes held together by some nongenetic material, somewhat like a string of pearls. This viewpoint is not correct. The exact opposite may, in fact, be true. It is possible that most crossing over may occur by the breakage and reunion of the genetic molecules themselves. In relationship to a growing human awareness to subtlety, a relationship in some cases considered as a "crossing over" or bridging of watery depths, and in all cases as a crossing over of barriers, this could represent the separation and reunion of the World Parents, the vehicles for psycho-spiritual inheritance; the facets of inheritance then, in this case some of which may be viewpoints, do not necessarily change as whole and interacting units but as units within which conceptual breakage and a reunion of hopefully more suitably organized constituents occurs. It is certainly difficult to imagine a change between viewpoints that is whole, one whole concept changed to another whole concept. It is not difficult to imagine, however, an exchange between facets of whole concepts.

Gene mapping describes whole concepts for genetic statements. The genetic fine structure of a number of viral and bacterial genes has been extensively mapped. The maps, showing varying lengths between genes, suggest that some gene products are larger than others. Given the variety of tasks assigned to genes this seems natural and expected. The expectation has proved true.

It has been shown that crossing over can cause recombination of the material within a gene, whatever the size of the gene may be. Genes contain many different but strictly linear sites where mutations occur and between which crossing over occurs. This fact has provided a powerful method for the investigation of the topological structure of genes.

There is the assumption that the structure of genes has a compatability with the structure of the form producing genes. That form takes on different shapes. It was first elucidated as the β-helix. In his typical style, James Watson describes a portion of his personal history as it helped lead toward the conclusion that the form producing genes is helical:

> "Luckily, merely a superficial grasp was needed to see why the TMV X-ray picture suggested a helix with a turn every 23 Å along the helical axis [in regard to protein subunits] . . . Francis [Crick] . . . maintained that the evidence was only so-so. My morale automatically went down, until I hit upon a foolproof reason why subunits should be helically arranged. In a moment of after-supper boredom I had read a Faraday Society Discussion on 'The Structure of Metals.' It contained an ingenious theory by the theoretician F. C. Frank on how crystals grow. Every time the calculations were properly done, the paradoxical answer emerged that the crystals could not grow at anywhere near the observed rates. Frank saw that the paradox vanished if crystals were not as regular as suspected, but contained dislocations resulting in the perpetual presence of cozy corners into which new molecules could fit."[12]

Watson was riding on a bus several days later when, en route to Oxford, relaxed contemplation allowed important realizations about these "cozy corners." He considered the TMV particle. It occurred to him that the particle could be like a tiny crystal growing in a fashion typical of other crystals, through the possession of "cozy corners."

"Most important, the simplest way to generate cozy corners was to have the subunits helically arranged. The idea was so simple that it had to be right. Every helical staircase I saw that weekend in Oxford made me more confident that other biological structures would also have helical symmetry. For over a week I pored over electron micrographs of muscle and collagen fibers, looking for hints of helices."[13]

As stated, the helix is the least of the four kinds of spiral. Spirals describe change. Helices describe a finitude befitting the material plane and the relatively tense and solid forms of organisms, but within this context, of course, change does exist, variability exists. The lyrical symbolism of the spiral describes the conditions of energy, the circulation of energy, magnetism and conductivity, nor in the helix are the beauty and benefit of symmetry and conductivity denied. Paul Calvert describes the background generating interest in the conductivity of DNA:

"For some time people have been looking for polymers with metallic conductivity or with superconductivity. . . . For a material to be conducting there must be a supply of charge carriers, electrons or 'holes,' and an easy, high mobility path for them to follow through the material . . . stacked planar aromatic molecules can provide an easy conduction path perpendicular to the molecular plane as in graphite. Hence the interest which developed at one time in the conductivity properties of DNA, where the stacking of the base pairs suggested that the molecule could be a 'one-dimensional' conductor. . . . Perhaps the most intriguing aspect of this work is that it requires cooperation between theoretical chemists, physicists, polymer chemists and engineers."[14]

Such cooperation would most certainly be required if the global structure of the conductive "ley lines" were to be extensively researched. *Ley line* is a term originating in British archaeology, for the straight magnetic pathways stretching across the earth at or just below its surface, on which are located sacred shrines and monuments of greater antiquity. It has been speculated that the global ley line structure reflects the principles of the tetrahedron, first of the five regular Platonic solids, which generates the spiral. The statement that "in our bodies is repeated the structure of the Whole" may not necessarily seem to bear out with respect to all of the specifics contained within that whole, at least as it is materially expressed; however, structural and other analogies are sufficient to suggest otherwise. How the global system of magnetic pathways finds literal analogy here remains to be seen; loose analogy to biological and psychic magnetics may be imagined.

The practical structural function of the double helix is to support hereditary factors. What is the function of those factors? A gene influences a particular set of characters and in a particular way. The functions of genes are, therefore, as multitudinous as such specificity implies. How is it that such units of function exist in DNA? Each gene belonging to the large group of structural genes is responsible for the exact sequence of amino acids building a specific polypeptide chain. This responsibility is expressed during protein synthesis. The set of characters influenced by a structural gene is that which depends, directly or indirectly, on the presence of its particular protein. Regulatory genes are productive by influencing the activities of the structural genes. DNA comprises a number of different kinds of genes.

Between the turn of the century and the early 1940s DNA was thought to be a molecule of comparatively diminutive size and to be composed of four small nucleotides. It was considered to possess marginal biological importance. In the reality defined by the current perspective on it, DNA forms one of the largest naturally occurring molecules and is composed of many thousands of nucleotides. Its significance is questioned only in relation to the mysteries of physics and the potential of the mind to influence the body.

The gene is replacing blood as the symbolic bond between parent and sibling. The heredity and evolution of individuals and species is affected by a factor that has itself evolved. The initial rules of genetic organization and evolution were revealed by studies of hemoglobin. The concept of gene families is a familiar product of these studies. A clue to the evolution of gene families is found in the evidence supplied by chromosomal mapping in which the genes of a given family form a line of developmental order, from genes expressed in the early embryo to genes expressed in the adult.

A portrait of globin genes, one sketched with recombinant DNA techniques, depicts a state of rapid evolutionary change undergone by dynamic gene families. The families are themselves unexpectedly large. Their constituents are specialized genes functioning at different developmental stages, genes that have apparently lost their function, genes acquiring new functions, and genes currently experiencing deletion and duplication in human and animal populations. The gene experience is one in which temporary change is effected by and upon these cellular constituents.

The evolution of genetic systems, including the genetic mechanisms and controls operative in a species, such as mutation, recombination, regulation, as well as other components directly related to the genetic material, exhibits a major function in terms of its relationship to factors providing a compromise between fitness, or the adaptedness requiring genetic constancy, and flexibility, or the adaptability requiring genetic variability. A reduced number of chromosomes is one of the factors that can influence the amount of genotypic diversity in a species. It has been characteristic throughout the history of evolution of a great many animal species-groups, such as *Drosophila,* and for many plants. Centric fusions have been responsible for these numerical reductions. Consequently, there has been little or no loss of genetic material in these cases. Genes that have assorted independently become segregated into different gametes after crossing over, not before. This has the result that fewer gametes contain recombinant genotypes in species, and there are fewer chromosomes in the genome. The reduction in the number of alleles in gametes provides that fewer kinds of genotypes will be generated in each generation. The equivalence of diversity in the gene pool of a species with between three and six linkage groups, as in *Drosophila,* does not alter this.

An example regarding the evolution of a cellular constituent is found in cytochrome c. This respiratory protein inhabiting aerobic prokaryotes and eukaryotes originated perhaps 1.2–1.4 billion years ago in an unknown prokaryotic ancestor of eukaryotes. The rates of evolutionary changes in protein are estimable from correlations between amino acid changes in particular proteins and the times various organisms have diverged according to fossil records.

This particular field is very new. There is insufficient evidence to properly judge many of its proposals. Biological evolution is not explained by simply viewing amino acid substitutions in proteins. Molecular phylogenies, however, provide information useful in combination with data obtained from fossil records and also from comparative studies of species currently living. The unsolved problems in this area are recognized as presenting some of the most exciting challenges in modern biology. Their resolution will very likely increase greatly the evolutionary understanding of the diversity of life.

Genes are no more stationary than the diversity of life forms they help to create. They are no more stationary than the cells they inhabit. Cells are in motion, a motion essential, as cell union and division, to the beginning and maintenance of life forms. While the cell is moving as a body, DNA is describing itself as a structure in motion; it is replicating.

Watson has written that the complementary shape of the double helix "immediately suggests self replication." This replication is the survival feature that must occur during the growth and reproduction of any living being. Its principle is simple: creation of the future by union of the past with the present, the old and the new. This is a familiar concept. Neumann has attributed a sacred place to self-creation (re-creation) in his *Origins and History of Consciousness*. The Buddhist theory of re-created events is also a sacred concept of activities necessarily enhancing the qualities and potential of a given life, and hence all life. The role of elderly persons traditionally fits this concept in terms of their functioning to unite the past, as well as its accumulated wisdom, with the potential of the future as it is exhibited in the present. If not for replication the double helix of DNA would stagnate.

The amount of DNA per diploid set of chromosomes is constant for a given organism. It is equal to twice the amount present in haploid sperm cells. DNA is also metabolically stable. It is not rapidly made. Nor is it broken down rapidly, as are many different cellular molecules. Atoms having become incorporated into DNA do not leave the DNA unless there is a break in the maintenance of healthy cell growth. These facts along with evidence from chemical studies with viruses and virus-infected cells provided early confirmation of the significance of DNA. A crucial experimental test came from isotopic study of the multiplication of T2. This subject is a virus containing a DNA core and a protective shell built up by the aggregation of different protein molecules.

The protein coat was labeled with radioactive isotope ^{35}S. The labeled virus was used for following the paths of phage protein and nucleic acid during the course of virus multiplication. The study was meant to elucidate which labeled atoms from the parental virus appeared in progeny of the phage. Parental nucleic acid was much in evidence in the progeny phage; none of the parental protein was detected. It became apparent that little of the parent protein ever enters the bacteria; it remains attached to the outside of the bacterial cell and has no discernible function after the DNA component has passed in. The primary function of viral protein is protection of its genetic nucleic acid component in its movement from cell to cell. When evidence against the assignment of a genetic role for the protein molecule was unavoidable the importance of DNA at last became apparent.

Two identical surfaces will not attract each other. It is far easier to visualize attraction between opposites, whether of oppositely shaped or oppositely charged groups. The double helix confounds this rule. Its two polynucleotide strands are complementary in both their shape and their charge. The replication of DNA involves strand separation and the formation of complementary molecules on each of the freed strands.

Each of the purine and pyrimidine bases can form several hydrogen bonds. The bonds are ideal forces for a template. They are highly specific. Their attractive forces are weaker and virtually independent of specific chemical groupings. Two molecules held together by several hydrogen bonds almost never appear in bonded form. The ratio of bonding to nonbonding in ordinary cellular conditions is about 10^4. The frequency with which adenine attaches to cytosine during DNA replication is about 10^{-8} times the frequency of its bonding to thymine ($10^{-4} \times 10^{-4}$). The G to C pair is held together by three hydrogen bonds. Replication is an event of greater precision.

The specificity involved in self-replication is included completely in DNA. This was evident in 1956 when DNA synthesis could be observed in a cell-free system prepared from the cells of *E. coli*. The DNA polymerase involved in the process was the first enzyme to be found in the synthesis of polydeoxynucleotides. It works only in the presence of DNA, which is necessary for the direction of the four nucleotides in the polynucleotide product. DNA polymerase cannot determine sequence specificity. It recognizes only the regular sugar-phosphate portion of the nucleotide precursors. Demonstration of this is possible by allowing the enzyme to operate in the presence of DNA molecules containing varying amounts of the base pairs of A-T and G-C. The enzymatically synthesized product will invariably have base ratios of primer DNA. No synthesis of protein or any other molecular class occurs during this cell-free synthesis. Compounds of non-DNA are eliminated as carriers of genetic specificity. The enzymatic product of DNA formation, like its primer, has a double helical structure. There is, as Watson reminds us, no doubt that *DNA is the direct template for its own formation.*

The actual process of replication appears to be very complicated. Clarification of the process was assisted by the discovery of a function of DNA polymerase. DNA polymerase is one of the major enzymes responsible in the linking up of nucleotides during DNA replication in *E. coli.* The assembly of complementary nucleotide sequences was evidently amenable to fairly straightforward test-tube analysis.

DNA replication involves strand separation. DNA molecules contain one heavy and one light strand after one generation of growth. They are thus of intermediate hybrid density. After two generations of growth half of the DNA molecules are light; the other half are hybrid.

Replication is the process which allows for a full set of chromosomes in the daughter cells of mitosis, cell division. In the spring of 1980 a model for replication was proposed by Drew M. Pardoll, Bert Vogelstein, and Donald S. Coffey, all of Johns Hopkins University. The complex of enzymes involved in replication was likened to the recording head of a tape recorder: the DNA "reels" through a fixed site. The proposed model was actually more complex than this analogy implies. The team stated that replication occurs simultaneously at many sites, known as replisomes, on a DNA molecule. Between every two sites of attachment, they suggested, DNA is brought in by the replisomes from both sides; newly produced strands form a loop. "This configuration," Coffey explained, "allows the cell to duplicate its DNA in an orderly manner without entanglement." Thus the nuclear matrix is implicated in the regulation of DNA synthesis in a way not previously considered, and replication may be bidirectional. Replication allows for, indeed creates, changes to which DNA accommodates for the sake of the survival of both itself and the organism in which it dwells.

"Purposeful movement is a universal attribute of life, so all-pervasive as to seem ordinary."—H. M. Schmeck, Jr.*

Cellular locomotion is one of dual significance, crucial to cellular life and to biological life as a whole. Three protein / keys affecting the capacity for cell movement are actin, myosin, and calmodulin. Techniques of cell propulsion include the amoeboid, ruffled membrane, flagelloid, and cilioid. The amoeboid technique, the means by which white cells travel the bloodstream they clean, is one in which the displacement of an interior substance promotes the movement of cells, and may result from a pull from the forward area within the cell, or perhaps a push from the rear of the cell. A hopping approach is described by the ruffled membrane technique visible in animal and human tissue cells. At each landing, or attachment, a ruffled membrane forms and the cells employing this technique move by attachment and detachment from the underlying surface, such as the glass of a laboratory dish, the membrane forming and reforming accordingly. With flagella cells are moved forward with the momentum of an ocean pushing a small boat. Finally, in another motion graceful only by virtue of our remove and because of the aquatic quality of the microscopic world of the cell, cilia rhythmically pulsate to propel cells, employing the cilioid technique just as similar waves clear the bronchial passages lined with cilia. But cells are no more in constant motion from place to place than are any other life forms. Typical of many meetings, certain cells have been seen to temporarily halt their growth and movement on encountering another cell. Behind action and inaction is the source of cellular energy, adenosine triphosphate (ATP). ATP allows for all the effects of cell locomotion, including wrinkles. On a microscopic level the wrinkles are no different than on the comparatively macro level. They are the wrinkles of stress. Here they express radial stress caused by the traction forces crawling cells exert in order to propel themselves outward.

ATP is a nucleotide coenzyme taking part in many chemical reactions in organisms. A wide range of, but not all, different cellular activities are dependent on ATP as a source of energy. ATP is formed from adenosine diphosphate (ADP), by utilization of light energy in

* "Cell Movement," *New York Times,* October 14, 1980.

photosynthesis and by means of the energy derived from catabolic processes. The sun appears to have reached into the depths of the hidden molecular world.

*COMMUNICATE: TO SHARE; TO PARTICIPATE**

Gene expression is the greatest event within cellular communication. It holds an ultimate responsibility for the nature of the phenotype. Three steps are fundamental to gene expression. These fundamental steps are replication, transcription, and translation.

Coded polypeptides essentially "underwrite" phenotypic development in an organism. The polypeptides arise from a process of transferral of information from DNA to create new DNA in replication and copying of DNA information into RNA during transcription; the latter information guides the synthesis of coded polypeptides during translation. Replication, then, is the synthesis of a replica sequence identical to the deoxyribonucleotides directing that process. Transcription is described by the synthesis of a complementary sequence of ribonucleotides in RNA, the sequence being complementary to the deoxyribonucleic sequencing which directs that synthesis. Translation occurs when the sequence of ribonucleotides in messenger RNA guides the synthesis of a corresponding sequence of amino acids in polypeptides.

DNA generally remains fixed in its location in the cell nucleus. Protein synthesis occurs in the surrounding cytoplasm. Proteins contribute vitally to cellular activities in many different ways. An intermediary is necessary for transportation of DNA instructions to the sites of protein synthesis. The logical candidate for such an intermediary or messenger is RNA. RNA is a polynucleotide. It is able to interact with DNA through complementary base pairing.

Three kinds of RNA interact during protein synthesis. These are messenger, ribosomal, and transfer RNAs: mRNA, rRNA, and tRNA, respectively. These are complementary copies of DNA sequences synthesized during transcription. Translation of genetic information into polypeptides is dependent upon them. Complementary basepairing determines the accuracy of interactions leading to accurate

* *Webster's Twentieth Century Dictionary,* Unabridged, Second Edition.

translation of an exact copy of the genetic instructions to be carried out.

Gene expression requires a flow of information from DNA-coded sequences to messenger RNA by transcription, and to proteins by translation of mRNA. Regulation of gene expression produces multicellular organisms emanating from a single fertilized egg cell. Every cell in an organism contains a complete set of genes. Differentiated gene action is responsible for phenotypic differences. Gene expression is generally controlled at the level of the transcription of genetic information into mRNA.

The operon concept of the regulation of gene action was formulated by the brilliant reasoning of Jacob and Monod. The operon is the result of operator mutations. Operator and repressor genes interact to control structural gene activity. Operator mutations define a gene cluster of closely linked DNA sequences acting as a coordinated unit of function. This unit is the operon. A policy of one operon, one mRNA exists. Several systems show that individual operons serve as the template for transcription of a specific type of mRNA molecule. Operons may not be the kings of cellular communication but inversion of just one letter provides a biological similarity of images between the operon and Oberon, a communicator in *A Midsummer Night's Dream.*

In the depths of biological organization communication is from cell to cell, and between the cell and the body of the organism. Various individual kinds of cells and systems of cells are responsible for operating the biological complex.

In the intellectual seat of the biological complex, the brain, a specialized system of nerve cells, neurons. The system functions to receive and provide instructions by means of electrical impulses directed over specific pathways. Understanding of the working of brain cells has remained at the primitive stage. Knowledge is reasonably adequate to a restricted task of describing and partially explaining how neurons generate and transmit electrical impulses. The impulses form the basic code element of the internal communication system in humans. The impulses arising at special receptor sites and taken to higher centers in the nervous system are collected and relayed by sensory neurons. Motor neurons carry impulses from the higher centers to "working" cells, such as muscle cells.

At the beginning of differentiation in the embryo, the body of a nerve cell, motor or sensory, sends out a long fiber, an axon, which grows toward its proper peripheral station to make contact with muscle or skin. Less than 0.001 inch thick, the adult human axon may be several feet long. It forms a miniature cable for conducting messages between the periphery and the central terminus inside the spinal canal or the skull.

Controversy as to whether or not the fundamental concept of the existence of individual cell units could be applied to the nervous system and its functional connections was not settled convincingly until the advent of high-resolution electron microscopy. Most of the surface of a nerve cell, including its extensions, is indeed closely invested and enveloped with other cells. The cytoplasm of adjacent cells, however, remains separated by distinct membranes. A small extracellular gap, usually of 100–200 Å, exists between adjoining cell membranes. Only a fraction of cell contacts are functional synapses, points of signal transference from one cell to the "next link in the chain."

Experimental methods of physiology have been more successful in dealing with the immediate processes of nerve communication than with more intractable long-term events. Comparatively little is known about chemical interactions between nerve and satellite. The same is true for the forces that guide and attract growing nerves along specific pathways and induce the formation of synaptic contacts with other cells. Nor is much known about how cells store information and provide memory.

Electrochemical events underlying the nerve impulse, action potential, have been greatly clarified since the middle 1900s. The voltage difference across the membrane is determined largely by differential permeability of the membrane to sodium and potassium ions. The nerve membrane is distinctive in its permeability. This permeability is regulated by the voltage difference across the membrane. This peculiar mutual influence is the basis of the signaling process.

How do cells receive stimuli? In most multicellular animals response to external and regulation of internal environments is enabled by specialized receptor cells sensitive to a wide variety of physical, chemical, and mechanical stimuli. The structure and function of receptor cells is diverse. The ways in which receptor cells connect with nerve fibers channeling into the central nervous system also vary. A common function, however, is of nerve impulse generation. The receptor cell creates a discharge of nerve impulses following mechanical deformation, light or heat absorption, or stimulation by a particular molecule. The events in this have not all been traced. A final event common to all sensory reception is this generation of nerve impulses. Each type of receptor communicates to higher nerve centers through its own private set of nerve channels. This enables the animal to decode the various messages.

Depolarization performs an important role in cellular communication. Stimulation of the receptor cell gives rise to sustained local depolarization of the sensory nerve fiber. The fiber then generates a train of impulses. Some of the first direct evidence for generator potentials at the cellular level was produced in 1935. H. K. Hartline, then at the Johnson Research Foundation of the University of Pennsylvania, found what appeared to be evidence of generator potential when he recorded the activity of a single optic nerve fiber and its receptor in the compound eye of *Limulus*. A train of nerve impulses was superimposed on the generator potential.

In 1950 Bernhard Katz, University College, London, obtained unmistakable evidence for generator potential in the vertebrate muscle spindle. Stretching of the spindle resulted in a small, steady depolarization recorded in the nerve fiber emanating from the spindle. The role of generator potential was confirmed by Stephen W. Kuffler and Carlos Eyzaguirre, then at Johns Hopkins University. Using the Alexandroqicz stretch receptor cells in crustaceans, they distorted the cell's dendrites by stretching the receptor muscle. The cell body became depolarized. The depolarization spread passively to the site of impulse generation in the axon. When generator potential reached a critical level, the cell fired a train of nerve impulses. The greater the depolarization of the axon above this critical level, the higher the discharge frequency. A receptor cell, then, triggers a release of nerve impulses by local depolarization of adjacent nerve fiber, whether its own fiber or that provided by another cell.

Direct experimental evidence of initial molecular events in the receptor process is exhibited in photosensory cells. Visual receptor cells in both vertebrates and invertebrates possess specially differentiated organelles containing a photosensitive pigment. Visual pigments are complex proteins. The light-absorbing part of the pigment, the chromophore, is the relatively simple vitamin A aldehyde. Containing numerous double chemical bonds, vitamin A aldehyde can exist in various molecular configurations referred to as *cis* and *trans* isomers.

Flicker fusion is a characteristic directly attributable to the generator potential, which becomes smooth at the highest repitition rates. A flicker is detectable as fluctuations in the generator potential giving rise to bursts of impulses. Flicker fusion is one of the more amusing aspects of cell communication. Motion pictures and television exhibit a light flickering which at a high rate appears not to be flickering at all. A neural basis for the phenomenon is apparent in generator potentials and nerve impulses recorded when, for instance, a *Limulus ommatidium* is exposed to light flickering at various rates. As the rate of repetition increases, the discharge rate becomes steadier; finally it is indistinguishable from a response to continuous illumination.

The responses of complex sense organs are determined by properties fundamental to individual receptor cells. They are also determined by influences they exert on each other, and by control exerted on them by other organs. The activity of receptor cells as communicators is integrated into complex patterns of nervous activity enabling organisms to survive variety and change.

Communication is the essence of genetics. The main characteristic of this communication is the channeling of information within the internal environment of an organism, and between the internal and external environments. This exchange necessarily includes communication between the cellular and overall physiological environments of the organism. Communication with the internal is of the first order, vital to the manifestation of what is, in a sense, beyond yet encompassing itself, much as cosmic consciousness is said to embody and transcend the individual consciousnesses upon which its character depends in its relation to the interaction between the individual and the whole.

Communication strives to accomplish two things: initiation, and maintenance. Both communication and the energy supporting it have been the chief focus of the twentieth century. Their characteristics and availability have undergone tremendous transformations. Genetic engineering is one of those transformations. The human understanding and maintenance of desired qualities of human life, especially the genetic complex, may, and its various components certainly do, face re-education.

An example of exemplary communication is found in the immune system. Immunity may be said to operate by asking questions. It challenges authority to prevent the dominance of possibly misguided, misunderstood, and unhealthy instruction to the genotype, instruction that could undermine if not the whole genetic complex directly, at least the well being of the physiological complex and thus of the genotype of the organism. While genetic engineering would not be possible given a genetic complex operating solely or largely as does immunity, organisms would not survive without the curious and challenging maturity of the immune system. Lack of a challenge would undermine not only immunity but also the whole system containing it, which then reverts to a vulnerable primal state as evidenced by afflicted individuals maintained in synthetic embryonic sacks. To see the tragedy of children physiologically afflicted by the vulnerable condition conferred by a lack of a properly functioning immune system is unnerving. Terrifying would be the sight of children intellectually afflicted by a formal educational system determining, as was refuted in the 1960s and suggested again in the early 1980s, that they be prevented from all the information and questioning by which their maturity necessarily evolves the balanced overview prerequisite to contributing to the well-being of the whole societal organism beyond yet encompassing self. There is a line between a healthy immune system and one that is lacking, is entirely absent, or is so excessive that it turns against its host as though the body were having an allergic reaction to itself. This line is no less seriously paralleled in the concerns of formal education.

Of the kinds of communication, psychic communication and its effect on the physiological and genetic complexes is least understood. Because it is so mysterious, operating in ways no existing tools have been able to adequately measure, curiosity has limited its frustration most commonly by either expressing disrespect for psychic phenomena or ignoring it all together. A better understanding of the mechanisms of psychic communication may encourage trust. Joint research ventures spurred by the success of exchanges of patients between the classical medical profession and traditional psychic healers promise some conclusive results before 1990. For the first time tools for measuring the operations of psychic healing could be promised by the nature of their developmental stages in 1981.

"The electron . . . runs within the body in accordance with its character within the body; that is to say, in accordance with the general plan of the body, and this plan includes the mental state."—Alfred North Whitehead

Although healing centers abound nationally and can be found internationally, and quackery exists among the various approaches to psychic healing just as it has throughout the history of classical medicine, many approaches and centers are quite valid. The independent and collaborative work of the Healing Light Center in California has contributed significantly to the creation of psychic healing as a valid practice and to an understanding of its mechanisms. The language of the psychic plane has, almost surprisingly for its presumed importance, remained virtually indiscernible to science while the language of a perhaps more materialistically natured mode of communication, through the gene, has with a certain suddenness become unmysterious.

The turbulent and transproductive 1960s provided many insights. In genetics the last of the 64 triplet codons (coding units) of DNA were deciphered in 1967. Four principal approaches have yielded the majority of information concerning the nature of the code:

1. Analysis of primary structures of wildtype and mutant proteins.
2. Analyses based on use of synthetic ribopolynucleotides to stimulate amino acid incorporation in cell-free systems.
3. Analysis of mutually enhancing binding to ribosomes of specific tRNAs and of synthetic mRNAs of different chain sizes and base sequences.
4. Inference from various genetical experiments, especially involving location and activity of mutations.

The code has been revealed as triplet. Each word, a codon, consists of three nucleotides whose instructions are punctuated to express the beginning and end of each genetic instruction. Each codon is specified by amino acids most of which qualify more than one of the triplets. The triplets are synonyms for all of the 20 included amino acids except methionine and tryptophan, which involve a single codon. The code is consistent and universal to nearly all life forms. The startling common bond implied by the genetic code reflects the equanimity with which diversity is unified. While other forms of heredity exist, such as extranuclear material, the gene is the prime mechanism of heredity. The functions of the gene are all the more valued in that they include the operation of the organism the gene has itself been instrumental in creating.

Primary gene action consists of transcribing DNA's coded genetic information into the ribopolynucleotide sequence of mRNA, and the translation, then, of this information into specific amino acid sequences. The four-letter alphabet of RNA, A, U (uracil), G, and C, is arranged into adjacent, nonoverlapping, three-letter codons. The codons dictate the position of specific amino acids in a polypeptide chain.

The action of acridines, such as the proflavin molecule, reveals these molecules as powerful mutagens that stretch and stiffen the backbone structure of DNA, intercalating, it is believed, between the base pairs of the DNA double helix. The stretching achieved by acridines encourages mutations:deletion or addition of one or more base pairs. The triplet code thus affected by acridines is read sequentially, in a reading frame. The double helix of DNA formulating the code consists of purines and pyrimidines bonded by a sugar and phosphate. Among the characteristics of the code, amusing until their effects are perceived in the phenotype, are nonsense and missense mutations, nonsense mutations producing incomplete polypeptide chains.

The two long, thin polymeric chains of the double helix are each polynucleotides, collections of nucleotides. The sugars of adjacent nucleotides are each linked together by a phosphate group. Ten nucleotides exist on each chain at every turn of the helix, every 34 Å. The distance per nucleotide base is 3.4 Å. The helix is approximately 20 Å. Four main nucleotides exist. They each contain a deoxyribose residue, a phosphate group, and a purine or pyrimidine base. The two pyrimidines are thymine (T) and cytosine (C). The two purines are adenine (A) and guanine (G). In RNA thymine is replaced by uracil.

The two strands of the double helix are joined by hydrogen bonds between pairs of relatively water-insoluble purine and pyrimidine molecular bases. In the bases, the purine, adenine, always pairs with the pyrimidine, thymine; and guanine with cytosine. Bonding directly between two purines would occupy too much space for a regular helix; bonding between two pyrimidines too little space. Opposites here definitely and critically attract. Within the bases, the two members of each couple are held together by hydrogen bonds. Adenine attached to thymine requires two hydrogen bonds, and guanine to cytosine requires three.

Hydrogen is an interesting gaseous element. It occurs most abundantly in combination with oxygen in the form of water. Water is the symbol of the Mysteries, of baptism, and initiation. In some ways, to survive contact with these, not immersion in but rather a bridging of the waters is necessary. The insolubility of purincs and pyrimidines express one approach to the dangers of dissolution.

The genetic code and code-letter combinations are described above in Tables 2–3 and 2–4.

Table 2-3. The Genetic Code.[a]

SECOND BASE A:

	Third Base A	G	T	C
A	phe[b]	phe	leu	leu
G	leu	leu	leu	leu
T	ileu	ileu	ileu	ileu
C	val	val	val	val

SECOND BASE G:

	Third Base A	G	T	C
A	ser	ser	ser	ser
G	pro	pro	pro	pro
T	thr	thr	thr	thr
C	ala	ala	ala	ala

SECOND BASE T:

	Third Base A	G	T	C
A	tyr	tyr	och[c]	amb[c]
G	his	his	gln	gln
T	asn	asn	lys	lys
C	asp	asp	glu	glu

SECOND BASE C:

	Third Base A	G	T	C
A	cys	cys	end[c]	trp
G	arg	arg	arg	arg
T	ser	ser	ser	ser
C	gly	gly	gly	gly

[a] After the Office of Technology Assessment.

[b] Symbology: ala, alanine; arg, arginine; asn, asparagine; asp, aspartic acid; asx, asn and/or asp; cys, cysteine; gln, glutamine; glu, glutamic acid; glx, gln and/or glu; gly, glycine; his, histidine; ileu, isoleucine; leu, leucine; lys, lysine; met, methionine; phe, phenylalanine; pro, proline; ser, serine; thr, threonine; trp, tryptophan; tyr, tyrosine; val, valine.

[c] Och (ochre), amb (amert), and end are step signals for translation, that is, they signal the end of synthesis of the protein chain.

Table 2-4. Code-Letter Combinations.[a]

Singlet Code (4 words)	Doublet Code (16 words)				Triplet Code (64 words)[b]			
					AAA	AAG	AAC	AAU
					AGA	AGG	AGC	AGU
					ACA	ACG	ACC	ACU
					AUA	AUG	AUC	AUU
					GAA	GAG	GAC	GAU
					GGA	GGG	GGC	GGU
A	AA	AG	AC	AU	GCA	GCG	GCC	GCU
G	GA	GG	GC	GU	GUA	GUG	GUC	GUU
C	CA	CG	CC	CU	CAA	CAG	CAC	CAU
U	UA	UG	UC	UU	CGA	CGG	CGC	CGU
					CCA	CCG	CCC	CCU
					CUA	CUG	CUC	CUU
					UAA	UAG	UAC	UII
					UGA	UGG	UGC	UGU
					UCA	UCG	UCC	UCU
					UUA	UUG	UUC	UUU

[a] Source: Nirenberg, Marshall W., "The Genetic Code: II," *Scientific American,* March 1963.

[b] Letter combinations increase with code word length. Given that 20 code words are needed to identify the 20 common amino acids, code length minimum is three letters, assuming all words of same length.

Fig. 1. A. van Leeuwenhoek, the first bacteriologist: frontispiece (portrait) and engraved titlepage. Antonius van Leeuwenhoek, *Arcana naturae detecta,* 1695, (By courtesy of the Wellcome Trustees, The Wellcome Institute for the History of Medicine.)

Fig. 1. *(Continued)*

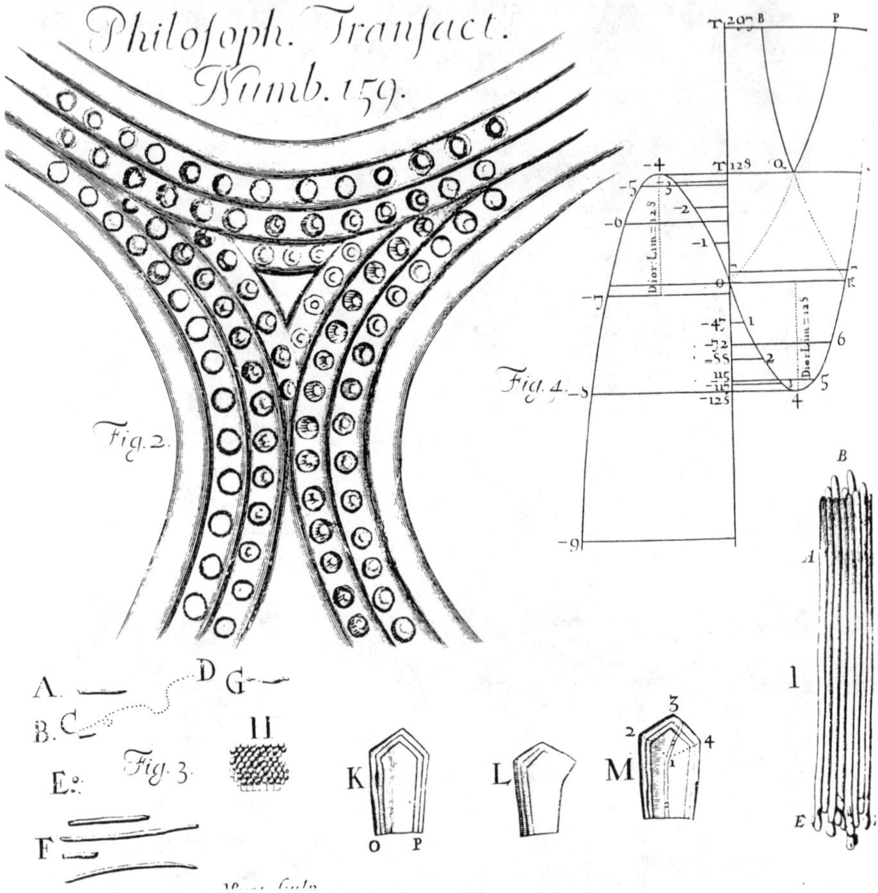

Fig. 2. Bacteria as van Leeuwenhoek visualized them through a primitive microscope and then drew them from memory. (From The Royal Society of London.)

Fig. 3. Josef Gottlieb Kölreuter, whose work was foundational to that of Mendel. (From the Generallandesarchiv.)

Fig. 4. The Altbrünn Monastery where Mendel lived and worked. Left, 1910 statue of Mendel. Center, rooms above arch are those which had been occupied by Mendel. (By permission of the Moravian Museum in Brno, Czecheslovakia.)

Fig. 5. Gregor Mendel, 1862. (By permission of the Moravian Museum, Brno.)

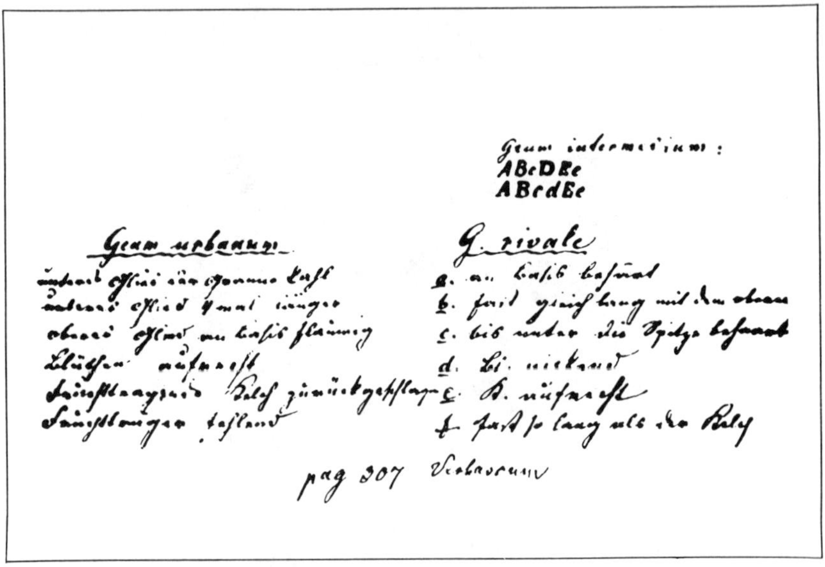

Geum intermedium:
ABcDEe
ABcdEe

Geum urbanum

G. rivale

Fig. 6. Mendel's notes on *Geum urbanum, rivale* and *intermedium;* and *pisum* Mendel's notes: a) *Geum urbanum* X *G. rivale,* contrasting characteristics, such as flowers erect *(G. urbanum),* flowers nodding *(G. rivale).* These notes were made in the flyleaf of Mendel's cope of Gaertner's book, *Versuche und Beobachtungen Uber die Bastarderzeugung im Pflanzenreich.* b) *Pisum.* Mendel also made these notes in his cope of Gaertner's book. Here Mendel is in the midst of conceptualizing his theory. He is hunting for clearly-marked differences for characters in various kinds of pea. Dunn, in his *A Short History of Genetics,* suggests that these notes on *Pisum* may well have been written by Mendel even before he pursued actual testing on 34 varieties of peas in 1854. Two of the characters described in his notes, i.e. seed form and pod shape, were chosen for experimentation and reported in 1865. (By permission of the Moravian Museum, Brno. Originals for Figures 4, 5, 6 are preserved in the Mendelium of the Moravian Museum.)

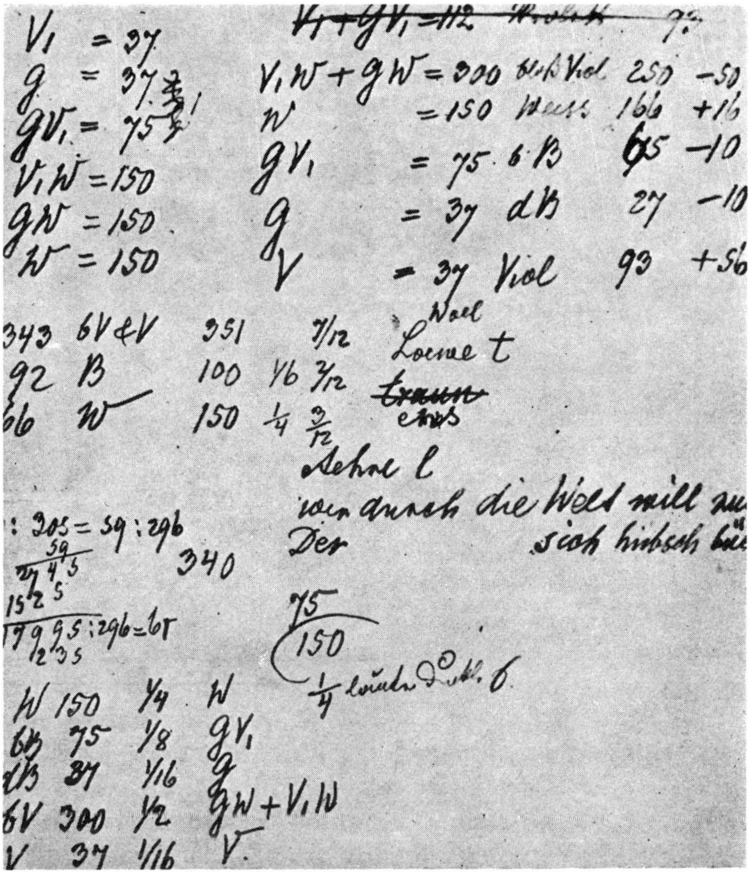

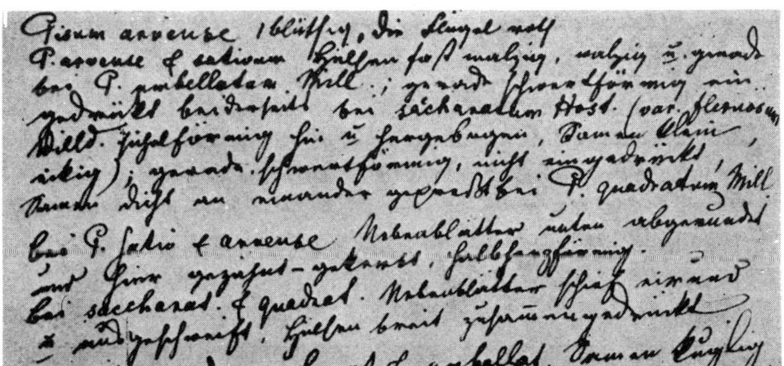

Fig. 6. *(Continued)*

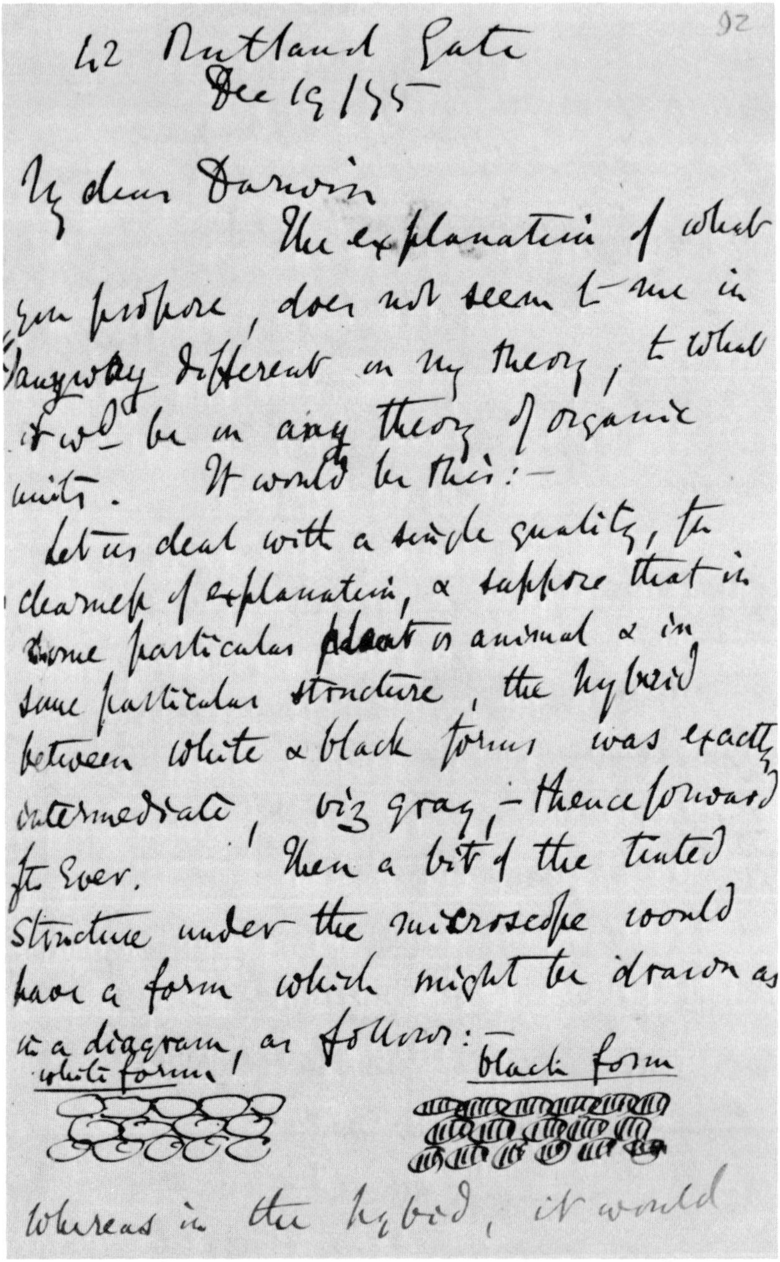
42 Rutland Gate
Dec 19/75

My dear Darwin

The explanation of what you propose, does not seem to me in any way different to my theory, to what it wd be on any theory of organic units. It would be this:—

Let us deal with a single quality, for clearness of explanation, & suppose that in some particular ~~plant~~ or animal & in some particular structure, the hybrid between white & black forms was exactly intermediate, viz gray, — thence forward for ever. Then a bit of the tinted structure under the microscope would have a form which might be drawn as in a diagram, as follows:—

white form

black form

Whereas in the hybrid, it would

Fig. 7. Galton's letter to Darwin, 1875, which contains a derivation of Mendelian ratios. (By permission of the Syndics of Cambridge University Library.)

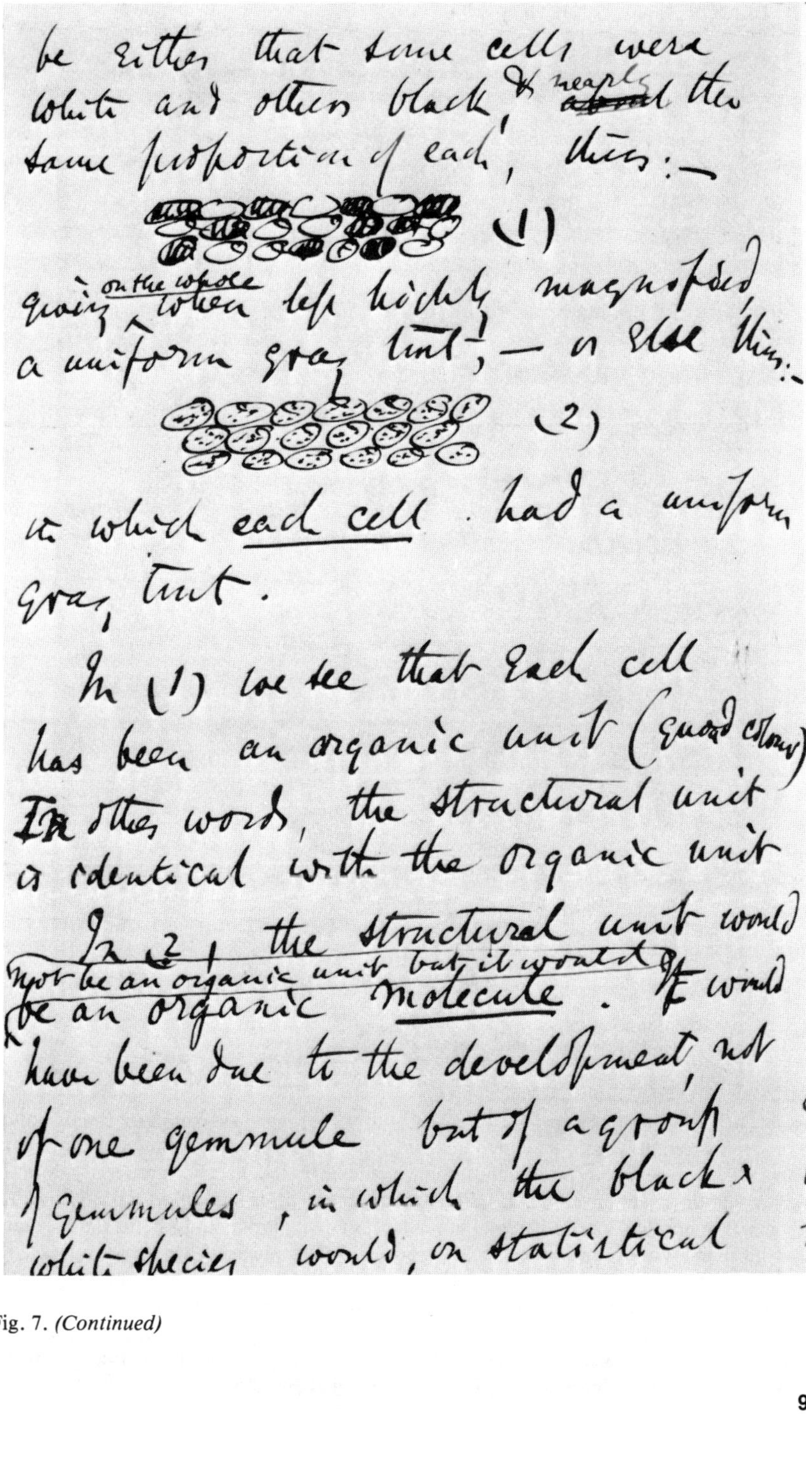
be either that some cells were
white and others black, & nearly the
same proportion of each, thus:—

(1)

giving on the whole when less highly magnified,
a uniform gray tint; — or else thus:—

(2)

in which each cell had a uniform
gray tint.

In (1) we see that each cell
has been an organic unit (quoad colour)
In other words, the structural unit
is identical with the organic unit

In 2, the structural unit would
not be an organic unit but it would
be an organic molecule. It would
have been due to the development, not
of one gemmule but of a group
of gemmules, in which the black &
white species would, on statistical

Fig. 7. *(Continued)*

Fig. 8. The Caduceus, by Holbein, 1523. The Caduceus has remained part of the insignia of the physician for countless centuries and to this day. It is otherwise known as the Staff of Mercury. Mercury separated the serpents of opposition with his staff; in this act he introduced a third element: equilibrium. The crowns Holbein shows above the serpents are the flames of wisdom. The serpents of opposition are fixed around the caduceus of equilibrium to form the figure handed down through history. The dove balances the power of the serpents. When Nature is tamed the opposition of the two forces expressed by the serpents manifests creativity (healing); when Nature is untamed this power is destructive. The current rendering of the caduceus, as a sign of the medical profession, gives two serpents entwined about the staff, which is capped by a globe on either side of which the wings are shown. One can imagine the serpents as the strands of DNA. (From Jill Purce, *The Mystic Spiral,* Fig. 11, Thames and Hudson.)

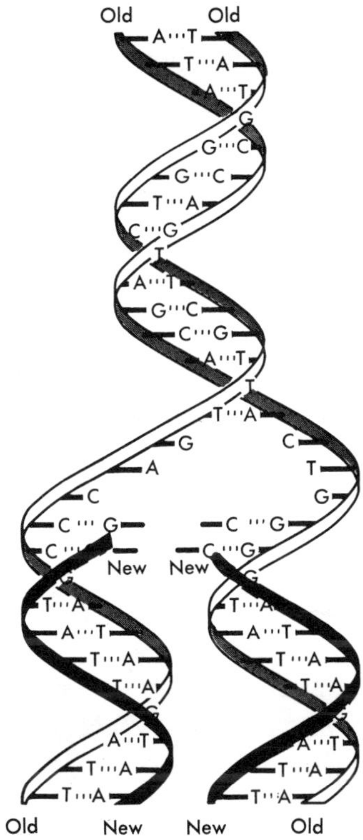

Fig. 9. DNA replication. Base pairs formed of purines and pyramidines: (A) Adenine. (C) Cytosine. (G) Guanine. (T) Thymine. (Courtesy James Watson, *Molecular Biology of the Gene,* W. A. Benjamin Inc., New York. Copyright 1970 by James Watson. Illustrations by Keith Roberts.)

Fig. 10. Watson and Crick in front of the model of DNA. Photo A. C. Barrington Brown. (From J. D. Watson, *The Double Helix,* Atheneum, New York, p. 215. Copyright 1968 by J. D. Watson.)

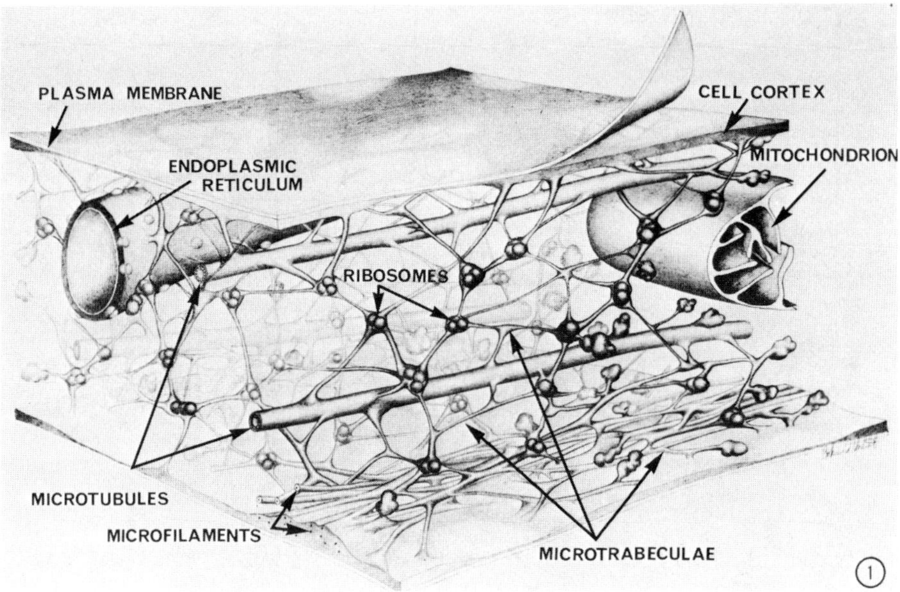

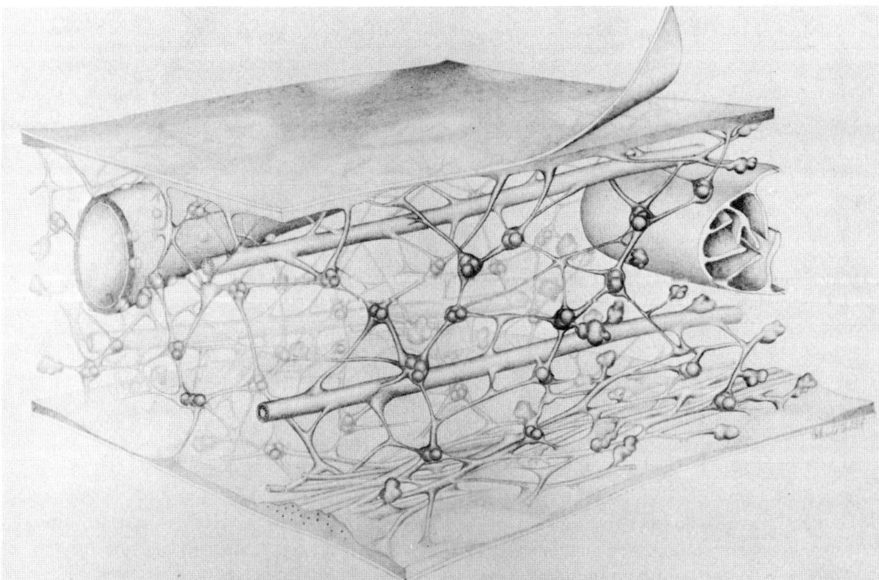

Fig. 11. Model showing a view ("Porter's view") of the space between the surfaces of a thinly spread cell. Important as physiological props at top and bottom are two cylindrical microtubules (pipe-like). The microtubules are enclosed by a lattice of micro-filaments (which are the basis of cell "crawl"); ribosomes are enmeshed in the filaments. At left, part of a mitochondrion is visible. By permission of K. R. Porter, and as it appears in *Inside The Cell: The New Frontier of Medical Science,* p. 64, Maya Pines, National Institutes of Health, 1978.

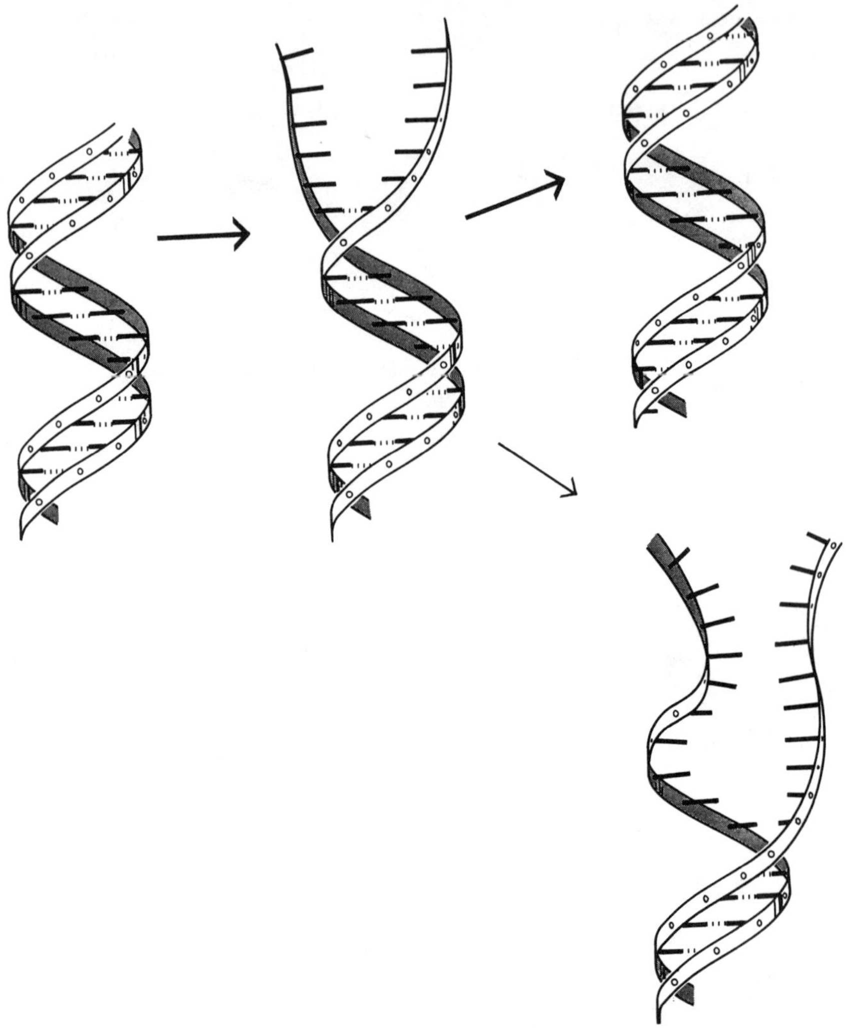

Fig. 12. Hydrogen-bonded base pairs in DNA. Two hydrogen bonds attach adenine to thymine. Three hydrogen bonds attach guanine to cytosine. (Courtesy James Watson, *Molecular Biology of the Gene,* W. A. Benjamin Inc., New York. Copyright 1970 by James Watson. Illustrations by Keith Roberts.)

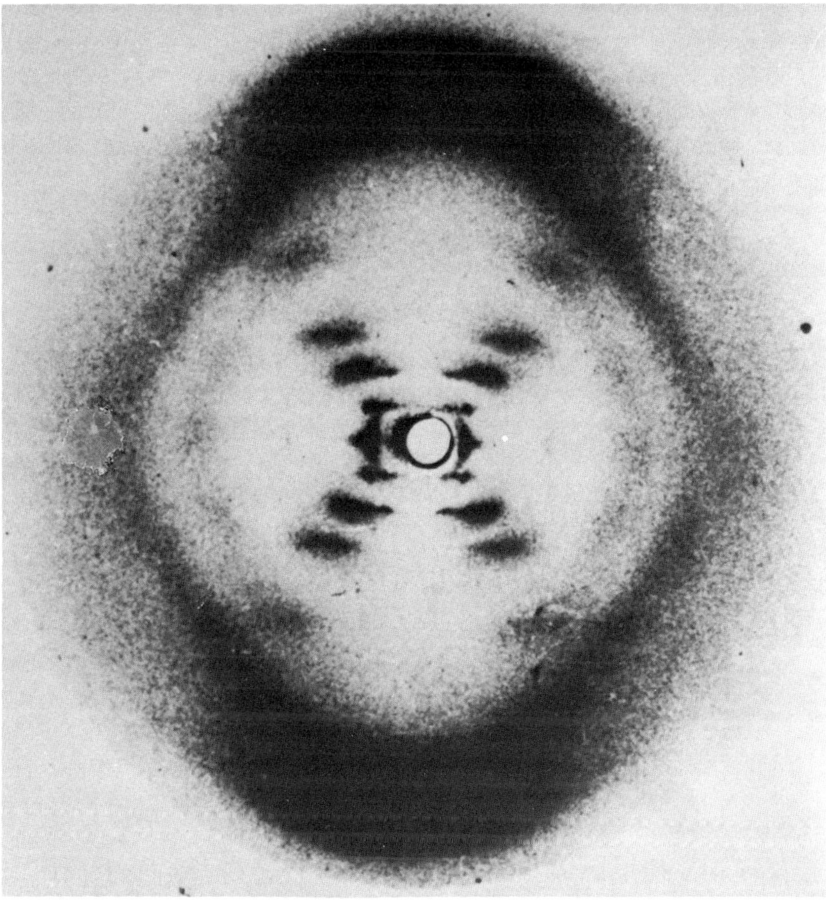

Fig. 13. X-ray diffraction photograph of DNA, B form, taken by Rosalind Franklin late in 1952. Revealed the general double helical structure of DNA to Watson and Crick. (From J. D. Watson, *The Double Helix,* Atheneum, New York, p. 168. © Copyright 1968 by J. D. Watson.)

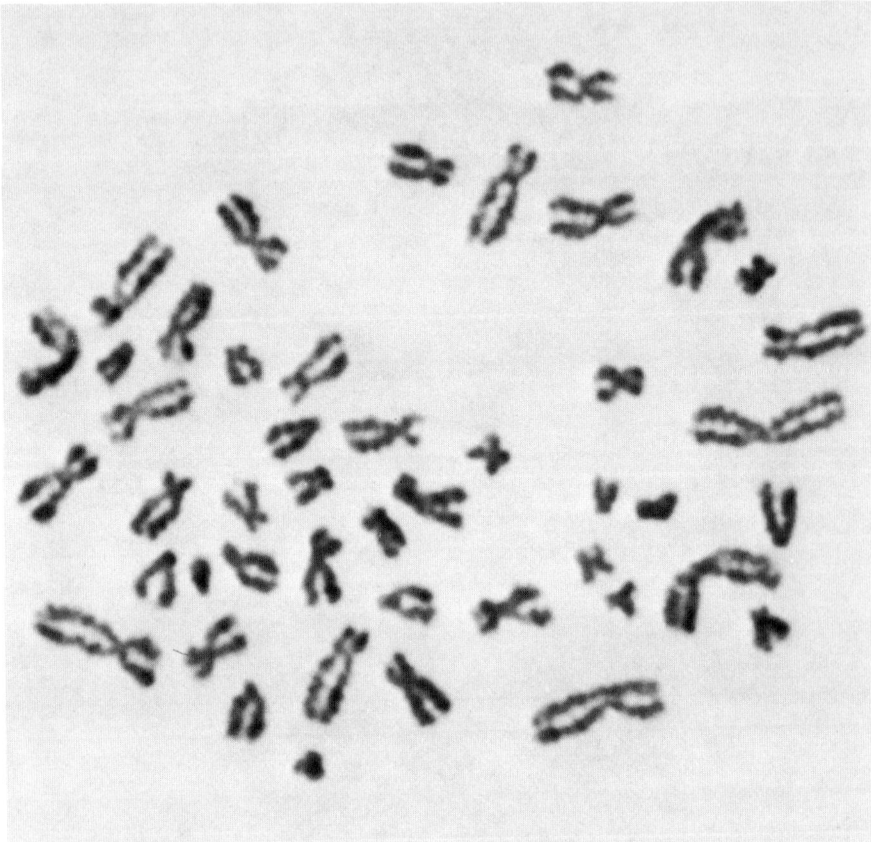

Fig. 14. Chromosomes in a normal human cell. The historic photo with which Joe Hin Tjio and Albert Levan showed that 46 individual chromosomes exist in a normal human cell, of which half are maternal, half paternal. Magnification: X2,300. (From Joe Hin Tjio, Department of Health and Human Services, National Institutes of Health.)

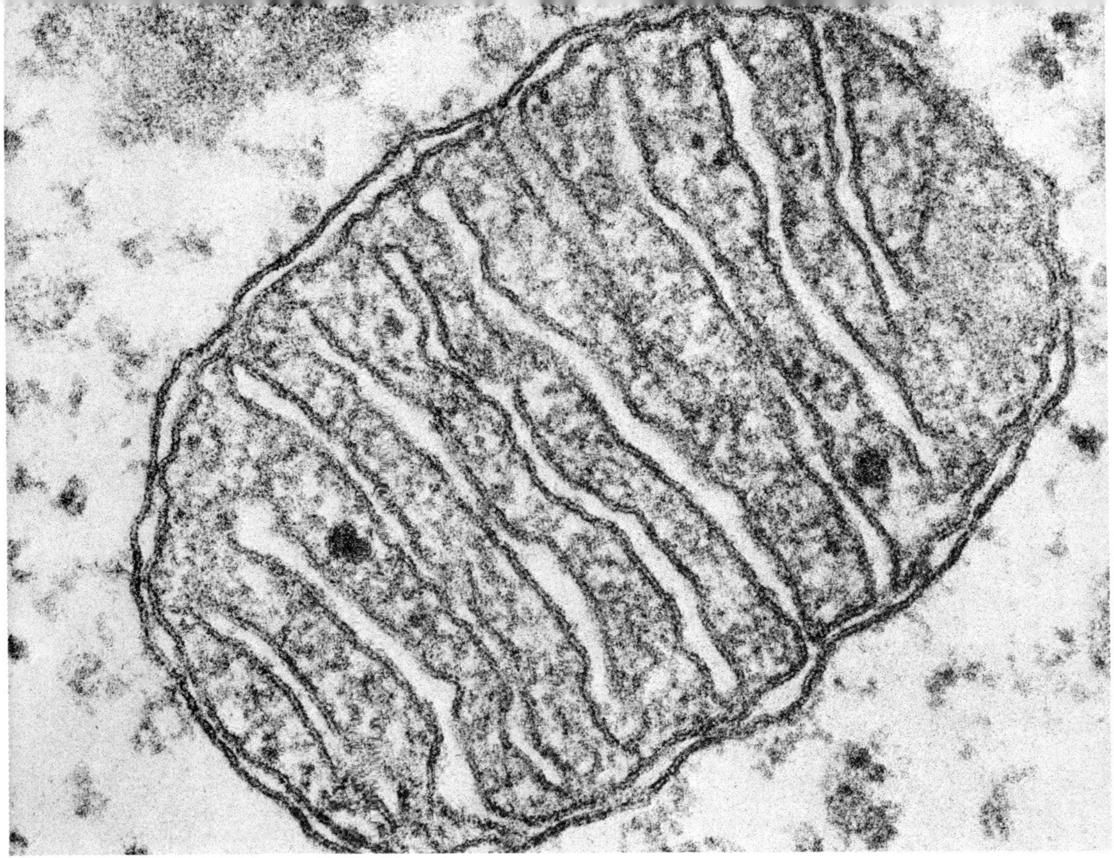

Fig. 15. Mitochondrion. Mitochondria contain their own DNA, and are major sources of cellular energy in organisms such as humans. Details of the single mitochondrion shown here are made clear by use of a high-powered electron microscope. Magnification: X228,000. (Courtesy of S G. E. Palade, Yale University Medical School.)

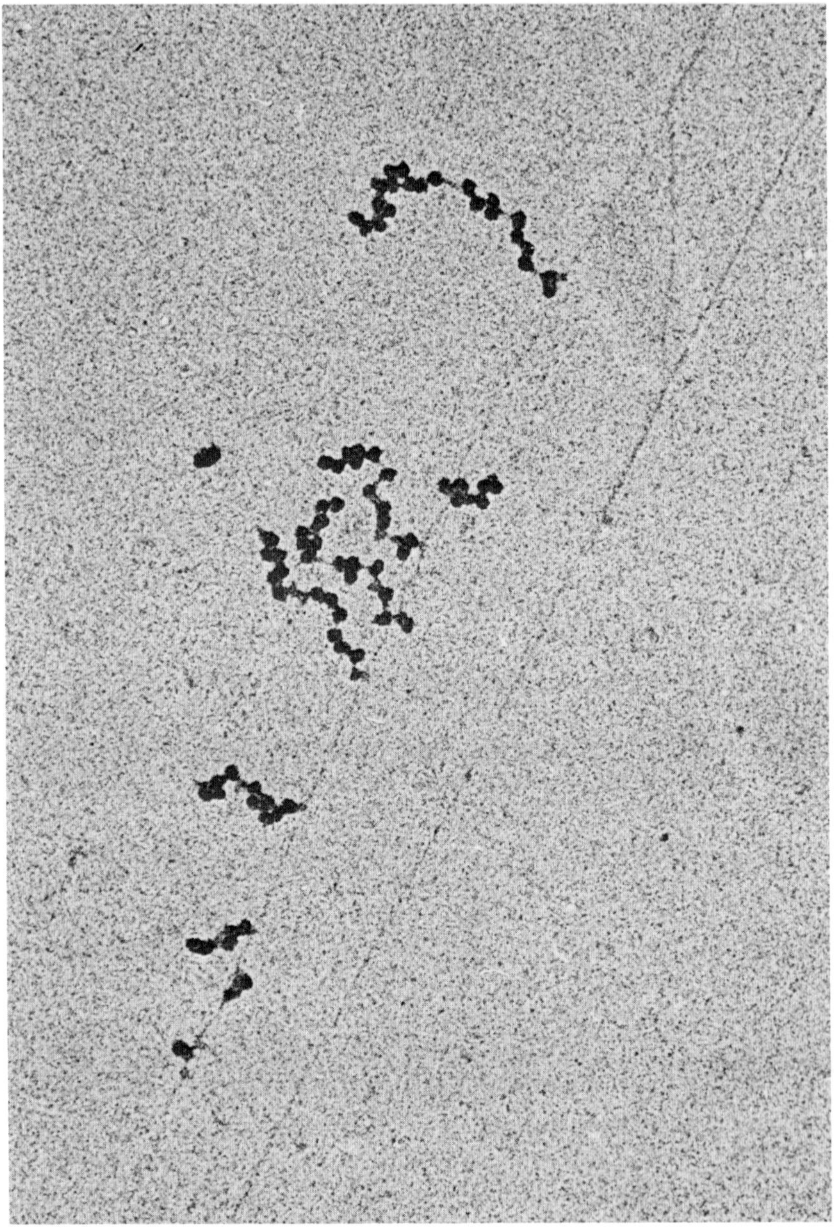

Fig. 16. 165,000X, An active unidentified operon of *Escherichia coli* gradient with attached polyribosomes. (Kindly supplied by O. L. Miller, Jr. Miller, O. L., Jr., Barbara A. Hamkalo and C. A. Thomas, Jr., *Science,* 169: 392–395, 1970, copyright is 1970 by the American Association for the Advancement of Science.)

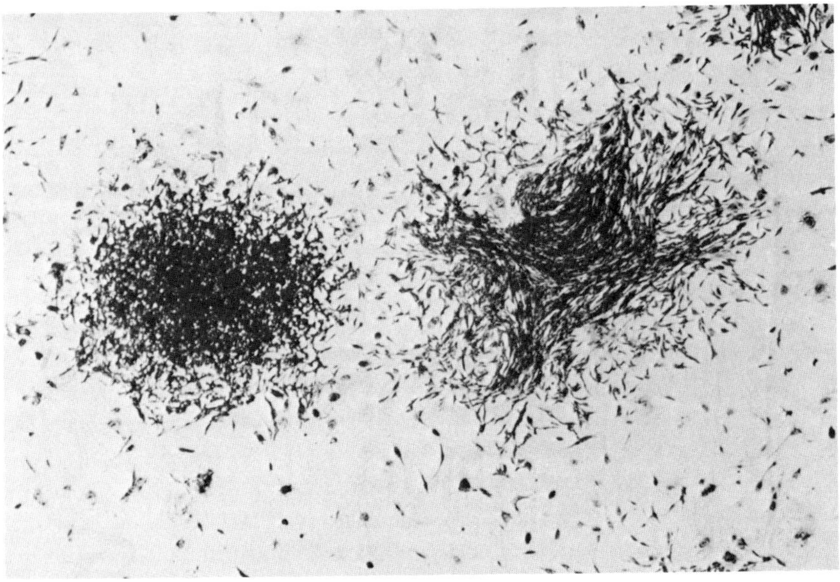

Fig. 17. Clones formed by transformed and normal hamster embryo cells. The transformed cells (left) are the result of the multiplication of a single cell in tissue culture. A virus enters. It transforms one of the tissue culture cells. As can be seen here, this cell has multiplied in a disorganized fashion. The transformed cells pile on top of each other with an apparent lack of contact inhibition. This creates a thicker (darker) mass than that found with the normal cells. (From M. Stoker, Cambridge.)

Fig. 18. The cover illustration of *Nature* giving mythological illustration to selfish DNA. (The illustration shows a lefthanded helix; this should be a righthanded helix.) (Courtesy W. F. Doolittle; *Nature,* Vol. 284, No. 5757, April 17, 1980. Illustration by Linda Angeloff Sapienza).

Section B
Applying Genetics

3
Promises and Realities

The discovery of DNA symbolizes the wonder, curiosity, and hope so familiar to humankind. The helical structure of DNA itself symbolizes the material complement to the mystic spiral.

"Although often intricate in form, the labyrinth is a spiral, and one which returns. It is a representation of the cosmos and all cosmoses, and hence of all ordered entities which correspond on the descending scale of analogy. It is therefore, at once the cosmos, the world, the individual life, the temple, the town, man, the womb—or intestines—of the Mother (earth), the convolutions of the brain, the consciousness, the heart, the pilgrimage, the journey, and the way."[1]

Along this journey are transformations of conditions. Among these transformations are health/illness, which transform each other, union/separateness, also transforming the nature of each other, and certain transformations of forms which promise to cope with all three categories of transformation . . . transcending any subdivisions we may attribute to form; here forms intermingle. Overall life is a compound consisting of multiple characteristics, unique and potentially conflicting forms, elements, and qualities; like the concept of worlds within worlds, there are compounds within compounds.

Unexpected or unusual compounds, such as those familiar to chemistry and now known also through genetic engineering, are not entirely new. Compound organisms date far into antiquity; special consideration is due compounds such as those attributed to the centaur, Chiron, to whom the art of military surgery is traced. New to the history of civilization are the unveiling dimensions of biotechnology including not only finely bred chimeras but also synthethics, or more appropriately, purified active principles, such as insulin and vaccines; manipulative abilities such as the evolving gene surgery, gene therapy, and gene transplantation are new; the extent to which knowledge is growing in regard to modifiable behavior and the mechanisms of aging is also new.

Chimeras are now evident from chemistry through biology. United with our technology we ourselves become chimeric, combining what is with what might be. In terms of genetic disease it has been said that the destinies of afflicted people will be remade by biotechnology. The extent to which this statement applies beyond genetic disease is vast. Depending on the materialistic relationship to consciousness, individual and cosmic, that vastness may be greater than is generally imagined.

The enzyme has become recognized as the tool by which gene splicing occurs in nature. The ability of the gene to be spliced is the most controversial and promising foundation of research endeavors in biology. The separation and insertion of sections of DNA propagated in host bacteria provide the basis of an education in the interdependent characteristics of viruses; when that action is applied in a particular way the result is a "new life form."

In the human being genetic engineering promises techniques for hereditary changes which give the meaning of the gene an intimacy resulting in reactions to the science similar to anxious reactions to an ordinary physician's office: horror and demand. Hugo de Vries foresaw the future when he wrote in 1901 that knowledge of the principles of mutation might enable the artificial induction of mutations; he further suggested in a 1904 lecture at the Station for Experimental Evolution at Cold Spring Harbor that penetrating roentgen rays be used in an attempt at altering the "hereditary particles" in germ cells.

The actual techniques of genetic engineering include transduction, conjugation, and recombination. These techniques are currently employed to alter "hereditary particles." Transduction requires viruses to alter the genetic material in bacteria. The infecting virus injects its DNA into the bacterial cell. New viral particles are formed.

During these formations chromosome fragments of the bacterium rather than the virul DNA may become incorporated into progeny viruses. The new viral particles, some of which contain bacterial chromosomal fragments, go on to infect a new cell. Recombination occurs between genetic elements incorporated from the first bacterium and homologous segments in the second.

Conjugation allows for the transfer of a plasmid during the mating of bacteria. The plasmid, an autonomous genetic element, can transfer the bacterial chromosome between two cells, so that in the recipient cell homologous segments of DNA can recombine. It should be pointed out, however, that only one type of plasmid can accomplish this; most plasmids transfer only themselves.

Recombinant DNA is the barrier-breaking technique allowing genes from one species to recombine with those from another. Restriction enzymes chemically cut DNA at specific sites and may occur naturally or by manipulation. Selected genes may be removed from donor molecules. The genes are inserted into plasmid DNA molecules. The resulting recombinant DNA may be propogated by cloning. Cloning affords large-scale production of a desired gene or protein.

The art of recombination has been automated more quickly than many arts. At roughly the same time Wall Street became interested in biotechnology the presence of automated processors began to be announced. Traditionally, every time a nucleotide is added to a chain under construction in solution, days can be necessary to separate the desired coding sequence. A 10-unit to 15-unit polynucleotide probes for a gene containing the desired sequence. Use of a probe either shorter or longer than this presents problems. A shorter probe might not pinpoint the desired gene but instead find many other genes containing the same, desired sequence. A longer probe is difficult to assemble. The solid-support method adapted from a technique used for the assembling of protein fragments, and making automated gene splicing feasible, reduces the complexity of gene-fragment assembly by firmly attaching a nucleotide to a solid particle much larger than itself; undesirables are simply flushed out. Since gene splicing is basic to genetic engineering and because the technology has come into such wide use, it is not surprising that machinery has been developed for the automation of research. That research has claimed a great amount of attention in the realm of pharmaceuticals.

The largest-selling prescription pharmaceuticals in the world between 1950 and 1980 were antimicrobial agents for treating infectious diseases. Most of these agents were antibiotics. Yet not a single antibiotic has had its complete biosynthetic pathway clearly elucidated. Antibodies giving immunity from infections are generated from sets of hundreds of genes. Their arrangements are some of the most elaborate yet discovered.

One of the most exciting and controversial developments undergoing further refinement is the ability to transfer genes. Genetic transfer was popularized in the late 1970s with the work of Michael Wigler, Richard Axel, and Saul Silverstein of the College of Physicians and Surgeons of Columbia University. Their object was the study of the control of gene expression in mammalian cells. Thymidine kinase (TK) was transplanted into mutant cells in the form of genes isolated from herpes simplex virus coding that enzyme. The transfer of genes allows for a multitude of biological changes including corrections such as this one with mutant cells deficient in TK. In other work, the rabbit gene for B-globin joined to a TK gene was transferred into mouse cells by Charles Weissman and his colleagues at the University of Zurich.

An early success in transfer between living animals (1980) involved a transfer of bone marrow cells between two sets of mice. The work, performed by Martin J. Cline, Winston Salser, Howard Stang, and Karen Mercola illustrated the introduction of drug resistance: the transferred gene was believed to confer resistance to the cancer therapeutic agent methotrexate (MTX). The direct potential here is the transfer of drug-resistance genes into cancer patients; the patients could then tolerate high doses of anticancer drugs.

Genetic transfer was a familiar concept as early as 1939. *Drosophila* became the recipient of DNA derived from the thymus glands of calves. The transplant was performed by Serge Gershenson, Zabolotny Institute of Microbiology and Virology, Academy of Science, Ukraine. In 1966 the first experimentally acceptable nonbacterial transfer was achieved: A. S. Fox and S. B. Yoon of Wisconsin University treated one strain of *Drosophila* with DNA extracted from a different strain.

Some of the problems in gene transfer can be likened to those of skin grafting, a procedure practiced as long ago as ancient Egypt, where cheek tissue filled the gaps left by the cutting weight of earrings in pierced ears. Will the graft or, in the case of genetics, the transfer be accepted? Furthermore, is genetic coding permanently altered by transplantation; in other words, are transferred genes inherited? What selective value has the transferred gene? Evidence along several lines supports the theory that foreign genes are integrated into chromosomes. Although foreign genes are not always favored by natural selection, which may view them as temporary or tending toward eventual recessivity, the fact that they are indeed sometimes favored is fuel to the controversy over biotechnological intervention in the gene pool, the collective of hereditary factors for a group or species.

As the 1980s began, the transplantation of purified genes by microinjection became an exciting reality in lower animals. Through the adaptation of a technique devised by Elaine Diacumakos, Rockefeller University, a genetic flaw in specific mouse cells was remedied. W. French Anderson, National Institutes of Health, and colleagues injected approximately 20 copies of a foreign gene into each cell. One experiment involved the incorporation of in vitro mouse cells with a viral gene producing an enzyme the cells had lacked. Another experiment resulted in the production of small amounts of a human blood protein by mouse cells. These are foundational experiments in the goal of genetic repair in human cells. (For an itemization of various possibilities in human genetic engineering see Table 3–1.)

If the gene is the mechanism of heredity and heredity participates in evolution, then evolution may be said to be undergoing repair, or at the very least, intervention. Studies of heredity and the hereditary mechanism have contributed surprises to the study of evolution, especially since the late 1970s. On the other hand, molecular anthropology finds little surprise in some of the statements coming from traditional paleontologists. A fact earthshaking to neither traditional nor molecular evolution students is that the globin molecule is one of "recent history." The announcement, however, that a distinctly human line seems to have diverged from the related australopithecine line more recently than expected by paleontologists naturally surprised members of this field; it did not surprise molecular anthropologist Vincent Sarich, University of California at Berkeley. Sarich found, with his colleagues, that molecular studies predicted just such a recent divergence. Mutation at the molecular level, states the basic tenet of molecular anthropology, occurs at a reasonably constant rate. An evolutionary clock appears.

New forms for analyzing biochemical fossils in terms of evolution

Table 3-1. Various Possible Technologies for Human Genetic Engineering.[a]

SELECTIVE MATING

By parental phenotype:
- Sterilization of "unfit."
- Encouraged mating of selected pairs.
- Artificial insemination with selected sperm donor.

By parental genotype:
- Follows the above but with specific knowledge of genotype.

By parental relationship:
- Avoidance of inbreeding, incest, etc.

By parental age:
- Encouraging of earlier pregnancy to avoid risk of chromosomal defects.

By genotype of gametes:
- Differentiation of sperm carrying "normal" gene from those carrying defective genes, followed by artificial insemination with preferred sperm.
- "Vaccination" of mother against defective sperm so that they are rejected by immune mechanisms.

TECHNOLOGIES INVOLVING DEVELOPING ZYGOTE

Parthenogenesis—development of unfertilized egg; results in genotype identical to mother.
Cloning.
Extracorporeal gestation:
- In vitro fertilization of egg; implantation in uterus of mother or surrogate mother.
- Egg already fertilized or developing embryo removed for development to term in artificial placenta.
- Completely in vitro birthing process—development of individual in virto from conception to birth.

TECHNOLOGIES INVOLVING SOMATIC CELL GENOTYPE

- Directed alterations of genes such as gene replacement or possibly specifically induced mutations.
- Random mutation with selection of cells with altered properties, followed by re-fusion of selected cells to host.

Those technologies best known to have realized their possibility, if not always their full potential, include sterilizing the "unfit," the encouraged mating of selected pairs attempted by eugenics, artificial insemination using a private donor or selection from one of the many sperm banks which have come into existence, in vitro fertilization, and directed alterations of genes, which is still in the experimental stage where, however, it has been successful.

[a] After Howard Brody, *Ethical Decisions in Medicine,* Little, Brown, 1981.

include analysis of distinctive biochemical pathways arriving at the same essential end product in different organisms, as well as examination of individual biochemical pathways which respond to different control mechanisms in diverse organisms. Possible ways to probe the past in these new ways have been well described by Philip E. Hartman and Sigmund R. Suskind (1969).[2] Studies of lysine pathways in a wide variety of species, as described by Hartman and Suskind, have depicted clear, consistent distributions which suggest that each pathway arose only once during evolution, a suggestion important to the concept of the uniqueness embracing and embraced by evolution. Hartman and Suskind suggest that one heme enzyme active in an important type of detoxication reaction and found in mammalian, avian, and reptilian livers but not in fishes or certain amphibians, may perform a significant role in the migration of animals from the marine to the terrestrial environment. These authors go on to state that, "Interestingly, there are many common reactions in the pathways leading to the formation of heme and chlorophyll. These homologies in reaction sequence again exemplify the types of correlations now available to the student of evolution."

The linking of studies in evolution with molecular studies is still new and is replete with speculation. The molecular evolutionary clock, for instance, considered in terms of rates of evolutionary change, past and present, remains something of a dream. It is a speculative concept. Comparison of DNA base composition, used as an additional criterion in modern taxonomy, is laborious. Closely related organisms bear very nearly similar DNA base compositions; organisms distantly related, however, share this characteristic and do not always differ significantly. Hybridization studies, rather than determination of nearest-neighbor or nucleotide frequencies, provide the potent tool. A single strand of a DNA molecule is able to hybridize with a complementary strand of either DNA or RNA. This annealing reaction assists determination of genetic relatedness.

Hartman and Suskind describe a future of synthetic messages, programmed cells, and programmed evolution:

"It takes little clairvoyance to realize that before long . . . refined techniques will result in the total synthesis of meaningful biologically active genetic messages. This can be predicted, for most of the predominant features of the genetic language are known, and the code appears to be universal. The fact that the methods for insertion and excision of genetic messages are routine procedures in microbial genetics assures us that the programming of metazoan cells hereditarily (using DNA messages) or transiently (using RNA messages) will not be far off . . . the possibility exists that human engineering with its attending moral and ethical problems will soon be a serious concern to scientists and to society in general. If somatic cells can be programmed and development altered (euphenics), then no doubt germ cells will be similarly amenable to change. Thus eugenics, or programmed evolution, may become an experimental reality. Many outstanding scientists envisage an impact of this on society even greater than that caused by developments in nuclear physics."

As the scenario unfolds significant discoveries are made. The raw material of evolution, for instance, may be what is commonly referred to as selfish DNA; if so, this form of DNA could certainly no longer be considered functionless. New relationships are formed routinely during the replication of DNA. They are also formed during the process of the study of DNA, both in our understanding of DNA and in the ways we strive to accomplish that understanding. Beside these new relationships are analogies and parallels of developments in different fields of study.

The parallels between developments in genetics and those in physics are strong and continuous. There are naturally areas of study in genetics which may, however, be relatively isolated, as well as those for which parallels are yet to be shown. Certainly, not all of the parallels involve a relationship between genetics and physics. As I have already mentioned, a parallel may be drawn between communication, in the structure of human society and in the immune system.

The development of new instruments for genetic analysis of immunology, promise what Leroy Hood, California Institute of Technology concludes will be equivalent to high-speed nuclear physics. The elusive role of the immune system in diabetics has already become more clear. This area has been cleared in numerous ways including the finding by University of Chicago researchers (1978) of diabetics exhibiting antibodies that damage the insulin-producing beta cells derived from rat pancreases. Other research has led to the identification of antibodies which may be important in diabetes. The disease has already received a blow from a process that makes insulin injections a more viable assistant to diabetics.

Discovery of the hormone insulin in 1921 assisted but did not solve the problems arising from diabetes. The art of feeding one's cells can be no less taxing than the preparation of a good meal for any other reason. Cell starvation, resulting from an inability to moderate blood-sugar levels in a natural fashion coupled with a reluctance to receive the necessary number of injections, remains uncontrolled for the majority of juvenile diabetics. Along with exploration of new methodology such as the infusion pump and artificial implantable pancreas is the improvement of the insulin commonly injected into diabetics. The foreign substance does not receive 100 percent acceptability among diabetics, because antibodies can readily form against it. The bacterial synthesis of human insulin may be a viable alternative to insulin derived from other creatures. Among others, Eli Lilly and Company announced in December, 1980, that its product had already gone into limited use.

Researchers, meanwhile, at the University of Kent in Great Britain, sought to replace many traditional synthetic sweeteners in diabetic diets. The protein thaumatin (talin), 3,000 times sweeter than cane sugar, was introduced into *E. coli* by way of the gene coding it. The bacteria then easily manufactured the sweetener, in a process preferable to that involving extraction from the West African fruit in which it naturally occurs in small amounts.

The most advanced applications of genetics in terms of technological sophistication and commercial development have been in the field of hormones. Genetic engineering has been used in attempts to prepare human peptide hormones including not only insulin but also growth hormone, among other hormones, with a view toward replacing extracts made from animal or human glands.

Hormones initiate and stimulate biological processes by bringing about the production of other hormones in other glands. They also act to maintain the organism's steady state, homeostasis. Certain hormones have attained a place in pharmacology, but their testing and use has been hindered by scarcity and costliness. Polypeptide hormones must be administered by injection, otherwise they are uselessly digested. Animal glands, human cadaver glands, and urine were, until recently, the only sources from which hormones could be drawn. A list of principal animal hormones, many of which are peptides, is shown in Table 3–2.

Table 3-2. Principal Animal Hormones.[a]

Hormone	Gross Effect of Hormone
Somatotropin (GH)	Growth hormone; stimulates growth and antagonizes action of insulin.
Adrenocorticotropic hormone (ACTH)	Stimulates secretion by adrenal cortex.
Follicle stimulating hormone	Stimulates ovarian follicle growth; estrogen secretion.
Luteinizing hormone (LH)	Stimulates ovulation.
Thyrotrophic hormone (TSH)	Stimulates thyroid secretions.
Insulin	Effect includes increasing glucose metabolism.
Glucagon	Effect includes antagonizing insulin action.
Thyroid hormone	Effect includes increasing metabolic rate.
Corticosterone	Effect includes antagonizing action of insulin on glucose metabolism.
Aldosterone	Operates in electrolyte balance.
Epinephrine	Glucose metabolism; vasocontrictor.
Estrogen	Growth of uterus; secondary sexual characteristics.
Progesterone	Maintenance of pregnancy.
Testosterone	Growth, secretion of male accessory glands.
Parathyroid hormone	Balance of calcium and phosphorus.
Antidiuretic hormone (vasopressin)	Increases water reabsorption in kidney tubules.
Oxytocin hormone	Causes contraction of smooth muscle; may assist in pregnancy termination.
Intermedin	Causes chromatophore changes and produces lightening and darkening of body color.
Gastrin	Causes secretion of stomach enzymes.
Secretin	Causes secretion of pancreatic HCO-3.
Pancreozymin	Causes secretion of pancreatic enzymes.
Cholecystokinin	Causes gallbladder contraction.

[a] Source: Office of Technology Assessment.

In addition to extraction, three technologies for production of polypeptide hormones and polypeptides have been developed:

- Chemical synthesis.
- Production by cells in tissue culture.
- Production by microbial fermentation after genetic engineering.

Large human polypeptides whose amino acid residues reach as high as 198 (prolactin) and molecular weight as high as 22,005 (growth hormone) and which are potentially attractive for biosynthesis include those shown in Table 3–3, while the smaller peptides are shown in Table 3–4.

In 1977 researchers at Genentech, Inc. inserted a totally synthetic DNA sequence into an *E. coli* plasmid. They demonstrated that the sequence led to the production of a 14-AA (amino acids) polypeptide sequence; that sequence corresponds to somatostatin, a hormone found in the brain. The experiment was possible because somatostatin's amino acid sequence was already largely known. The existence of sensitive assays allowed for detection of the hormone's expression.

Somatostatin is one of about 20 recognized small human polypep-

Table 3–3. Large Human Polypeptides of Potential Biosynthesis.[a]

Prolactin	Cholecystokinin (CCK-39)
Placental lactogen	Big gastrin (BG)
Growth hormone	Active fragment of PTH
Nerve growth factor	Cholecystokinin (CCK-33)
Parathyroid Hormone (PTH)	Calcitonin
Proinsulin	Endorphins
Insulinlike growth factors (IGF-1 and IGF-2)	Glucagon
Epidermal growth factor	Thymosin-α1
Insulin	Vasoactive intestinal peptide (VIP)
Thymopoietin	Secretin
Gastric inhibitory polypeptide (GIP)	Active fragment of ACTH
Corticotropin (CCK-3)	Motilin

Of these, seven were currently used in medical practice in 1981 (growth hormone, insulin, corticotropin (ACTH), calcitonin, glucagon, secretin, active fragment of ACTH).

[a] Source: Office of Technology Assessment.

Table 3-4. Small Peptides of Potential Medical Interest.[a]

Dynorphin	Caerulein
Little gastrin (LG)	Bradykinin
Somatostatin	Vasopressin (ADH)
Bombesin	Oxytocin
Melanocyte stimulating hormone	Facteur thymique serique (FTH)
Active dynorphin fragment	Substance P (4–11) octapeptide
Neurotensin	Angiotensin II and III
Mini-gastrin (G13)	MSH / ACTH 4–10
Luteinizing hormone-releasing hormone (LNRH)	Enkephalins
Active fragment of CCK	Active fragments of thymopoietin (TP5)
Angiotensin I	Thyrotropin releasing hormone (TRH)

Three of these, vasopressin (ADH), Oxytocin, and thyrotropin releasing hormone (TRH), were current to medical practice in 1981.

[a] Source: Office of Technology Assessment.

tides chemically synthesized without difficulty. The development of biosynthetic processes for production of small peptides looks especially to five areas: (1) the 7-AA sequence known as MSH / ACTH 4–10 reputed to influence memory, concentration, and other psychological-behavioral qualities; (2) cholecystokinin (33 AA) and bombesin (10 AA) appetite suppressors; (3) hormones, such as somatostatin, which are released by nerves in the hypothalamus of the brain to stimulate or inhibit the pituitary gland to release hormones; (4) calcitonin (32 AA), the largest polypeptide produced by chemical synthesis for commercial pharmaceutical use (as of 1981), which is useful for pathologic bone disorders; (5) adrenocorticotropic hormone (ACTH) (39 AA), which promotes and maintains normal growth and development of adrenal glands, and also stimulates the secretion of other hormones. This last, ACTH, is a diagnostic agent for adrenal insufficiency. In principle, ACTH could be used for many medical disorders, including rheumatic disorders, allergic states, and eye inflammation.

Polypeptides of high molecular weight cannot be made practically by chemical synthesis. They must be extracted from human or animal tissues, produced in cells growing in culture, or manufactured by fermentation using genetically designed bacteria. The last method has produced insulin and human growth hormone.

Composed of two chains of amino acids, insulin is initially produced as a single, long chain called pre-proinsulin. This is cut into a shorter chain, proinsulin. Proinsulin forms the A and B chains of insulin when a piece is cleaved from the middle. Rapid research developments have occurred in the genetic engineering of insulin. Colonies of *E. coli* excreting rat insulin were achieved only a year after the insulin gene had been incorporated into *E. coli* without the success of its expression there. Genentech, in collaboration with a group at City of Hope Medical Center, announced soon afterward the separate synthesis of the A (21 AA) and B (30 AA) chains of human insulin.

Animal and human insulin do not differ very much. Human and porcine insulins differ in a single amino acid. Human and cattle insulins differ with respect to three. These variations have not been shown to impair the effectiveness of insulin; testing of the use of human insulin in a diabetic population has, however, been limited. What is known is that a segment of this population responds adversely or produces antibodies to animal insulin, experiencing subsequent allergic or resistant reactions. Bioengineered production of human insulin, states the Office of Technology Assessment (OTA), may become a kind of insurance for diabetics.

Growth hormone, which is normally dependent on the presence of insulin, is another subject of interest in genetic engineering. The growth of an organism involves a highly complex process. The process depends on the correct balancing of variables. Growth hormone (GH) is one of a family of closely related, relatively large pituitary peptide hormones, single-chain polypeptides 191 to 198 AA in length. It induces growth in, among others, many soft tissues, cartilage, and bone. Growth hormone is required for postnatal growth in humans. Its action depends on the presence of insulin, whose secretion it stimulates. The biological significance of GH is clearly illustrated by the growth retardation which results from its absence before puberty; the benefits of the application of hGH in replacement therapy provide further evidence of its role in growth.

The National Pituitary Agency, created in the early 1960s to collect pituitary glands from coroners and private donors, satisfies most of the demand for human growth hormone (hGH) in the United States. The hormone has been provided without charge to treat children suffering from hypopituitarism and dwarfism. A special need for hGH has existed, however, for treatment of senile osteoporosis (bone decalcification), for other nonpituitary growth deficiencies such as

Turner's syndrome, for intrauterine growth retardation, for bleeding ulcers uncontrollable by other means, and for burn, wound, and bone-fracture healing. Synthesis of the hormone so that large quantities can be produced in bacteria has become a goal of the new genetics. Genentech represents one of two groups who had announced by 1981 the preparation of microorganisms with the capacity for synthesizing GH.

Molecular cloning (recombinant DNA, or rDNA) has aimed at attempts with other important polypeptide hormones, including:

- Parathyroid hormone (84 AA).
- Nerve growth factor (118 AA), which influences development, maintenance, and repair of nerve cells.
- Erythropoietin, a glycopeptide largely responsible for regulating blood cell development, which promises applications from hemorrhages and burns to anemias and other hematologic conditions.

The proteins of the immune system, antigens, interferons, cytokines, and antibodies, all promise to be affected by recombinant DNA technology. Polypeptides are at the very heart of immunology. Applied genetics will play a critical role in developing a pharmacology for controlling immunologic functions.

The immune system does not cover all possible fronts in the human environment; we therefore have vaccines to assist us in combating disease. For potential new vaccines see Table 3–5. Pure vaccines based on antigens synthesized by rDNA have been proposed to fight communicable diseases like malaria, diseases resistant to classical preventative efforts. Vaccines directed against toxic proteins, like diphtheria toxin produced by some organisms, prepare the body to neutralize them. Molecular cloning promises to produce inactivated toxins. Immunity conferred by live vaccines usually exceeds that conferred by nonliving antigenic material. Engineered microorganisms may be capable of becoming productive sources of high-potency antigen. They would offer far larger, more sustained doses of vaccine without side effects from the contaminants which may be found in vaccines consisting of killed microorganisms.

Influenza vaccines are among those with promising futures. The

Table 3-5. Major Disease for Which Vaccines Might be Developed[a]

PARASITIC DISEASES:	BACTERIAL DISEASES:
Hookworm	Dysentery
Trachoma	Typhoid fever
Malaria	Cholera
Schistomiasis (river blindness)	Traveler's Diarrhea
Sleeping sickness	
VIRUSES:	
Hepatitis	
Influenza	
Foot-and-mouth disease (for cloven-hoofed animals)	
Newcastle disease virus (for poultry)	
Herpes simplex	
Mumps	
Measles	
Common cold rhinoviruses	

[a] Source: Office of Technology Assessment.

cause of influenza is a virus that has been uncontrollable. This uncontrollability is largely because of the frequency with which the virus mutates and changes its antigenic structure. Antigenic protein genes for influenza could be kept in a "gene bank" for use as needed. The genetic code for several antigens could be introduced into an organism such as *E. coli* to create a vaccine with several antigens produced in one fermentation.

Approximately 800 million doses of foot-and-mouth virus (FMDV) vaccine are used annually worldwide. The virus is produced in larger volume than any other virus for vaccine purposes. Enormous quantities of the virus must be contained for production of the vaccine. But this containment refers not only to quantity but also to the housing of the material. "Many outbreaks [of the disease] are attributed to incompletely inactivated vaccine or to the escape of the virus from factories."[3] Molecular cloning could produce a more stable vaccine with less risk of escaped virus. Foot-and-mouth virus vaccine made by molecular cloning is expected to be distributed commercially by 1985.

Vaccines against viral hepatitis have also received significant attention. Hepatitis is reported to be second only to tuberculosis as a cause of death by infectious disease. Conceivably, rDNA technology may lead to other uses of antigens as well. Vaccination against parasites, including malarial and hookworm, immunization in connection with cancer treatment, and counteraction of abnormal antibodies creating autoimmune diseases such as multiple sclerosis are all possibilities.

One of the most tremendous bursts of excitement in modern genetics has been the result of studies in interferon. Glycoproteins produced in a cell in response to infection by a virus, interferons inhibit the multiplication of viruses. Preliminary studies support the use of interferon in treating viral diseases such as rabies, hepatitis, and various herpes infections. Extraction of interferon from leukocytes, white blood cells, has been the popular method of production. Tissue culture production as well as rDNA have provided techniques also widely used. Recombinant DNA is regarded as the key to mass production of interferons. At least four major biotechnology companies have been working on improved production methods for interferon.

This is perhaps a good place to remember how remarkable is the ability to perceive the gene and even directly and specifically affect a genome. This kind of specificity is an enormous leap from Mendel and hybridization.

When research in Switzerland and Japan led to the report (1980)[4] that the structure of genes for various kinds of human interferon had been deciphered tremendous excitement ensued. Understanding the structure of a gene for interferon allows for inference of the structure of the interferon itself. Interferon has undergone study not only as an antiviral agent but also as an anticancer agent. Genentech, in collaboration with Hoffman-LaRoche Inc., has produced two forms of human interferon using rDNA. Biogen S. A. has also produced interferon. As in other cases, technologically produced interferon is, or at least appears to be thus far as effective as naturally produced material. In 1981 researchers at Genentech revealed that instead of employing genetically engineered bacteria to produce interferon they had begun work with yeast. The yeast was better suited to high-volume production and put bacterial systems into the position of being merely "alternative" systems for producing human proteins. This statement describes a horizon not yet fully realized, for bacterial systems are still very much in use. However, this future use of yeasts, the potential for which arose in connection with interferon, is a rewarding coincidence for those involved in the production of interferon.

Interferon is a lymphokine. Lymphokines and cytokines are regulatory molecules referred to as glycoproteins. Lymphokines, biologically active soluble factors produced by white cells, have been impliated at virtually every stage in immune response. At least 100 different compounds are included. Cytokines, similar to lymphokines, include compounds associated with the thymus gland, thymic hormones. In 1979 the controversially named Biological Response Modifier (BRM) subcommittee of the National Cancer Institute (NCI) concluded that several of these agents may have significant potential for cancer treatment. Industrial production of glycoproteins has been stymied by want of a production system. Yeasts may be the most useful microorganism for their production. The problems of molecular cloning of glycoproteins present an area in need of greater attention.

Antibodies are the most extensively exploited protein components of the immune system. Between 1975 and 1980 antibodies produced from cells in culture allowed levels of purity never before attainable. The development of monoclonal antibodies is very much responsible for this high level of purity. Antibodies are particular about the antigens they will recognize. Monoclonal antibodies will recognize only one kind of antigen. Monoclonal antibodies began attracting attention as the result of work performed by Drs. Caesar Milstein and Georges Kohler, Medical Research Council, England. Milstein and Kohler fused two types of cells, myeloma and plasma-spleen cells. The resulting hybridomas produced monoclonal antibodies. The hybridomas can be grown in mass culture or as tumors in mice and a virtually limitless supply can thus be made available.

Between 1960 and 1980 segments of the diagnostic and clinical laboratory industries independently developed for the purposes of detection and quantification of particular substances in specimens. Hybridomas will doubtless replace animals as the source of antibodies used in diagnosis and monitoring. Their accuracy has been a boon to testing. Monoclonal antibodies assist in the detection of hormone levels, of the presence of proteins, and of the presence of drugs in blood, and in the monitoring of the blood or tissue level of a drug. They also identify microbial pathogens.

Applications of monoclonal antibodies also include improving the acceptance of organ transplants by injection of the recipient with antibodies against certain antigens. They may find application in a reversible immunological approach to contraception by action against an antigen involved in reproduction. The seemingly diverse range of application of monoclonal antibodies includes their possible applicability to tumor localization and also to the targeting of cancer cells with antibodies carrying anticancer chemicals.

At a late 1979 meeting (National Institute of Allergy and Infectious Diseases) on how molecular approaches such as genetic engineering could be applied in the development of vaccines, the conclusion was that the road to such an application of the new technology would be long and arduous. Progress would be hindered especially by limited understanding of the immune system. Yet, advances have been made. By 1980 the study of monoclonal antibodies had precipitated greater optimism. Monoclonal antibodies have helped to unveil many of the mysteries of disease and immunology. Seeking out and attaching to cellular and tissue sites, monoclonal antibodies are a key to mapping the location of drug effects on the brain and other organs. Research into multiple sclerosis using monoclonal antibodies permitted detection of chemistry changes in patients experiencing flare-ups of the disease and led to the discovery of a common deficiency of immunological suppressor T cells among these patients (Harvard Medical School, 1980). Monoclonal antibodies, which may be employed as a diagnostic tool that is used also for virus "fingerprinting" and for that cell fusion resulting in antibiotic hybridomas (hybrid cells), have been known to exist. The first practical way to obtain pure samples of monoclonal antibodies was reported by Kohler and Milstein (1975).

At a time when every facet of the cell seems remarkable and promising little is overlooked in the shuffle, least of all the value of proteins. The life span of individual proteins in living organisms is relatively short. Hemoglobin lives about four months; serum albumin only a week or two. Aged proteins are digested by proteolytic enzymes of tissues. A significant portion of recovered amino acids may be available for biosynthesis of new proteins; another portion is catabolized and the nitrogen excreted. From amino acids living organisms can synthesize their own proteins.

Structural proteins such as collagens, elastins, and keratins (the compounds of extracellular structures such as hair and connective tissue), albumins, globulins, and a variety of others, may be genetically engineered. Structural proteins, however, present difficult problems for molecular genetic manipulations. Their size and complexity exceed synthetic and analytic capabilities. Their medical use has yet to be established. Additionally, the material derived from animals has appeared quite adequate wherever there has been need of it. With so little encouragement the value of this particular research becomes questionable. Is it conceivable that research in one particular area of the hard sciences can be a candidate for indirect application to another area; could it even be a tool in the contemplation of scientific research aimed not simply at materialistic application but also at symbolic interpretation? The research of structural proteins does have at least one candidate for direct application: Serum albumin is a candidate for molecular cloning. It is the smallest of the main plasma proteins. Serum albumin is used for treating shock and blood loss. It has been the principal product of blood plasma fractionation.

Every cancer can be linked to a chromosome defect, proposes Jorge J. Yunis, Univesity of Minnesota. Cancer research is unyielding in its efforts to understand the disease. Cancer treatment, however, has remained largely traditionalized and can lead to removal of a part of the body of the patient, as when a masectomy is performed. The destruction, rather than the removal of substances such as tumors, presents an ultimate goal which many cancer patients may find desirable. Macrophages may achieve this destruction of cancer cells. Macrophages are scavenger defense cells. They attach themselves to cancer cells clustering around them; in this way the macrophages prepare their targets for chemical destruction by hydrogen peroxide, by enzymes, or by other substances produced by the macrophages. Research in this field has been extensively performed.

In 1979 the unchanging nature of cancer cells was called into question by Beatrice Mintz, Institute for Cancer Research, Fox Chase Cancer Center, Philadelphia. Mintz demonstrated that the deadly cancer cells of mice inserted into new embryos develop normal tissue growth and normal mice. The research carried out on succeeding generations shows that the cells stemming from those of a cancerous nature do not revert to malignancy. They do not fuse with other cells but, to use Mintz's words, "retain . . . genetic integrity." While it is rather difficult to understand what quality of the unique embryonic condition diverts cancer cells from their usual behavior, this research is hopeful. The research cannot yet be called promising, however, as the embryonic state is one of uniqueness and purity in which many otherwise expected results do not occur. Mintz showed that cancer cells could revert to normal under certain conditions. Approaches to curing cancer that evolve from these studies reach beyond existing cancer therapies in their hopes. A new beginning in the confrontation with cancer could be one in which therapy avoids the necessity of becoming a matter of killing cancer cells.

"The elixir of life and the fountain of youth are goals of biochemistry."
—Arthur Talbot Bawden (1943)[5]

In June 1980 an article titled "Immortality," in *Science and Living Tomorrow,* began with the statement that a hunt is on for DNA switches that tell the body to shift into the progressively lower gear known as aging. An intriguing question has been one which suggests the possibility that the mind is responsible for the initial triggering of the switch. It is well recognized that aging is not entirely biologically automatic but, at the very least, is a combination of nature (cellular characteristics) and nurture (societal characteristics, pressures). Herein lies a great opportunity for interactions between studies of the placebo effect, "healing" such as the laying on of hands, and mood and behavior, each of which is valid in terms of either psychology or physics (as the study of energy), and genetics, as well as more broadly scoped physiological study. However long an individual may desire to live, the possibility of living longer and more actively than the current means of medicine permits may be a possibility for which society is ill prepared. Unsolved problems, especially ethical ones, already abound in the concern over the existing prolongation of life, over abortion and when life begins, and over coma and when life ends.

A review of diverse cultures shows similar aging patterns. Age is commonly believed to be clearly associated with multisystem organ decay, which in turn renders the body more susceptible to the stresses of any particular life-style, to individual states of physiology, and to environmental factors. This is the typical attitude and it is one in which the possibility of exceptions to the rule becoming the rule are considered unrealistic. The medical view is justifiably aimed at what is and makes but limited comment as to what might be.

It is difficult to deny the view that in most cases death is more than simply tiring out and the ensuing complications. This allows for a conceptual disassociation of pure aging from death, a disassociation relatively easy to perceive when considering overall human vulnerability and extenuating circumstances such as behavior and environment for which further research precludes a definitive statement. Many causes of death are perceived in the general archetypal image of the Trickster, tester of our ability to cope with human vulnerability, although the same tester may be a benefactor of strength, a foe turned friend, as Lao Tze infers.

The search for the genetic aging mechanism has spurred intriguing research. As an example, research concerning the chromosomes of 14 strains of mice showed significant differences in the life span of mice genetically identical except at key locations in the major histocompatibility complex. While the research is not conclusive it does support the profound idea that, whether the mind only recognizes or actually triggers biochemical changes, genes control longevity. This idea was first suggested by Richard Cutler of the National Institute for Aging.

An attitude of assault, of being assaulted, by the common conditions of life is the popular view among individuals feeling that the living which ages them is a constant and burdensome challenge. This attitude is itself being assaulted, or at least encouraged to grow beyond itself. Lewis Thomas, Sloan Kettering Memorial, is one of the most staunch supporters of bacterial innocence. In 1980 and 1981 it became apparent that many attitudes toward illness and how it comes about were not justified. In terms of bacteria the cause of illness is not the invaders but the response of the recipient individual. Attitudes toward disease are necessarily changing with the results and suggestions of current medically oriented research. The changing attitudes affect the nature of that evolving research itself. A modulating attitude of centuries of medical attention is again shifting from what may be imagined as the original attitude of the healer. Hundreds of thousands of years ago initial appreciation of the healing art, writes Robert Silverberg,[6] "went something like this: the healer, undoubtedly considered a man of supernatural power, likely had his first patient in himself. A slashed finger may have become infected and puffy. It throbbed with pain. 'If I cut the finger off . . . the demon that possesses it will not be able to invade the rest of my body.'" This attitude has begun changing to one of: "My body can resist this demon. I am stronger than it is. I am strong enough to neutralize or transform this demon's influence. I can protect my body and still keep my finger." We are reaching for a point when exorcism is no longer necessary, making gods of ourselves and equating many of the promises of genetic engineering as rivaling God. It is our response to disease, not our transcendence of it, which leads us to these conclusions.

"What besides of sorrow, and dejection, and despair Our frailty can sustain."—Milton[7]

Genetic diseases include everything from Down's syndrome to general behavioral disorders to self-mutilation. The tracking and therefore potential prevention of genetic disease has been greatly assisted by molecular biology. A new approach to the problem initiated in the late 1970s by researchers at the Massachusetts Institute of Technology as well as at other prominent institutions, provided the basis for realistically aiming at the production of linkage maps tracing the inheritance of genetic disease through families. The study promises to be greatly assisted by the availability of genealogical records of the Mormon population, whose well-kept histories date to 1538 and earlier. This constitutes the largest human genetic study in the world. It is being conducted at the University of Utah, Department of Medical Genetics.

Genetic diseases have potential treatment in the future of genetic engineering. Genetic engineering is dependent upon enzymes. The term *enzyme* refers collectively to a group of proteins catalyzing various chemical reactions. Enzymes have been known since the early 1600s. Recognition of enzymes was encouraged by their role in digestion and fermentation processes which result in alcohols and other products. Approximately 1000 enzymes are known, 875 of which were listed and named in 1964 by the International Union of Biochemistry. Enzymes catalyze a staggering variety of chemical reactions. They are often complex, yet bring about relatively simple transformations.

Enzymes function as catalysts in reactions that involve the metabolism of living organisms. Their roles in biological chemistry and medicine are vital. Knowledge of the life processes and expansion of that knowledge depend critically upon the investigation of enzymes, their constitution, structure, synthesis, and behavior. The common properties of enzymes, their origin from living cells, their predominant, established role as catalysts for several types of chemical reactions, their structure suggesting that most enzymes are simple or conjugated proteins, depict a commonality that includes relatively high sensitivity to environmental conditions, temperature, and pH (acidity). The environmental tolerance of enzymes closely parallels that of other substances associated with life processes, for they tolerate a relatively narrow temperature span.

The great number of reactions catalyzed by enzymes in living organisms includes all the oxidation processes by which the organism obtains its energy, whether mechanical or thermal. It also includes hydrolysis processes by which food carbohydrates, proteins, and fats are broken down into simpler molecules for direct oxidation; enzymes also catalyze all the detoxification reactions by which many harmful substances that may be absorbed by the organism are converted for excretion.

Enzymes are involved in virtually every biological process. They are potent, versatile, and diverse. They perform, however, a small role in the practice of medicine. Conventional chemistry cannot synthesize enzymes. Commonly, enzymes are extracted from human blood, urine, or organs, or are produced by micro-enzymes. Exploitation of rDNA clones as a source of enzymes opens one of the few conceivable routes to inexpensive production of compatible human enzymes. Somatic cell fusion and culture are also routes toward this end.

Biotechnology demands a clinical need large enough to justify investment. As of 1981 hemophiliacs have received the only administration of enzymes, proenzymes converted to active enzymes in the body as needed. The enzyme urokinase, however, may be used to achieve almost the reverse case of hemophilia. Urokinase could also remove unwanted blood clots leading to strokes, myocardial infarcations, and pulmonary emboli. The drug is normally isolated from urine or produced in tissue culture. Urokinase is most popular outside the United States. The drug has not been as readily accepted here as by European and Japanese medical communities.

One of the greatest triumphs of microbiology has been with antibiotics. The first published account of microbial antagonism is William Roberts' report in the *Philosophical Transactions of the Royal Society,* 1874. Roberts noted that fungi growth often suppressed the growth of bacteria. He stated: "I have repeatedly observed that liquids in which the *Penicillium glaucum* was growing luxuriantly could with difficulty be artificially infected with bacteria." The observation specifically mentions the effect of a *Penicillium* on bacterial growth. It foreshadows the work of Fleming 50 years later. Yet, the account did not attract attention. It was forgotten until its rediscovery by medical historians nearly 100 years later. By 1958 Sheehan had synthesized 6-aminopenicillanic acid. The door opened for semisynthetic production of an antibiotic.

For three decades, from 1950 to the present, antimicrobial agents have been the largest-selling prescription pharmaceuticals in the world. Most of these were and are antibiotics. Antibiotics are antimicrobials naturally produced by microorganisms. Antibiotics provide most of the precedent for employment of microbial fermentation to produce useful medical substances. High-yielding antibiotic-producing microorganisms identified by their selection from among mutant strains. Organisms producing new antibiotics are isolated by broad screening efforts such as soil sampling. They are then cultured in the laboratory, where efforts are made toward the improvement of their productivity.

Antibiotics are usually nonprotein. They are complex substances, generally the end products of a series of biological steps. Knowledge of molecular details in metabolism has increased the understanding of antibiotics. Yet, elucidation of the complete biosynthetic pathways of antibiotics remains desirable. The limits of knowledge in the area of antibiotics does not entirely jeopardize manipulation. Mutations can be induced within original microorganisms so that production levels are increased.

Various methods can increase production of antibiotics. From among these methods the creation of new antibiotics may also emerge. Microbial mating, which leads to natural recombination, is an interchange that has been widely investigated as a way of developing producers of high-yielding antibiotics. The use of this natural breeding method has been limited by, among other things, mating incompatibility among many higher fungi of industrial importance. Natural recombination is most advantageous when strains of extremely diverse origins are mated. The proprietary secrets protecting commercial strains usually lead to the unavailability of the divergent "competing" strains most likely to produce the union of vigorous hybrids.

Protoplast or cell fusion provides a convenient method for establishing a recombinant system in those strains, species, and genera lacking efficient natural means for mating. The technique applies to nearly all antibiotic producers. It can combine benefits developed by mutation and selection in divergent lines. Research has included comparing the qualities of a fungus resulting from the mating of fungi where one fungus has been produced by protoplast fusion. The conclusion of the comparison between techniques, which involved *Cephalosporium acremonium,* was that protoplast fusion is far superior to mating in terms of the desired results.

Protoplast fusion which leads to the transference of entire sets of genes, not just one or a few genes, can cross the established barriers of genus and species. The preservation of traits and long-range stability of fused clones requires resolution for protoplast fusion to become a popular method for end-result-production of antibiotics. A greater problem has been determining what to look for and how to recognize it.

Many microorganisms cannot be induced to produce desirable quantities of antibiotics. Techniques in recombinant DNA are able to improve strains. Plasmids may be used to control the synthesis of certain antibiotics. Some plasmids may even nonspecifically enhance antibiotic production and excretion. With this technology there becomes feasible the transfer into a new host the whole group of genes necessary to produce an antibiotic. The genus of greatest interest for use is *Streptomyces.* Of the 2400 known antibiotics produced, 2000 are made with *Streptomyces.*

Organic chemical entities, except for antibiotics, have been extracted from natural plant or animal sources. They may also be synthesized chemically. Enzyme conversion of raw materials is not unfamiliar to pharmacology. This conversion method is common for steroid hormones.

In 1949 cortisone was found to be an agent useful to treating arthritis. The demand for it could not be met since no practical method for large-scale production existed. Chemical synthesis was complicated and expensive. The early and mid-1950's brought reports from many investigators on the microbial transformation of several intermediates to compounds corresponding to the scheme of chemical synthesis. Saving many chemical steps and achieving higher yields allowed manufacturers to make the drug more available. Processes inefficient in 1981 faced the possibility of future improvement by genetic techniques. Chemical substances such as those used to treat hypertension, as well as morphine and important vasodilators are produced by a series of enzymes coded by corresponding genes. The identification of genes important to this process will determine the feasibility of fermentation methods.

Genetics is the study of form and function on a microscopic scale critical to the physiological nature of larger forms and functions. The study of form, speaking generally:

> "... may be descriptive merely, or it may become analytical. We begin by describing the shape of an object in the simple words of common speech: we end by defining it in the precise language of mathematics; and the one method tends to follow the other in strict scientific order and historical continuity. Thus, for instance, the form of the earth, of a raindrop or a rainbow, the shape of the hanging chain, or the path of a stone thrown up into the air, may all be described, however, inadequately, in common words; but when we have learned to comprehend and to define the sphere, the catenary, or the parabola, we have made a wonderful and perhaps a manifold advance. The mathematical definition of a 'form' has a quality of precision which was quite lacking in our earlier stage of mere description; it is expressed in few words or in still briefer symbols, and these words or symbols are so pregnant with meaning that thought itself is economised; we are brought by means of it in touch with Galileo's aphorism (as old as Plato, as old as Pythagoras, as old perhaps as the wisdom of the Egyptians), that 'the Book of Nature' is written in characters of Geometry.'"[8]

Transformations and recombinations of form and alteration of the functions of form is what genetic engineering is about. It is also what chimeras, familiar to mythology and now to genetics, are about. The transformations and recombinations resulting in chimeras are concerned with parts, attributes; these parts affect the nature of whole entities. The problem of studying form in terms of its independent parts, or as though each of its parts were indeed independent, is a subject approached by D'Arcy Thompson:

"With the 'characters' of Mendelian genetics there is no fault to be found; tall and short, rough and smooth, plain or coloured are opposite tendencies or contrasting qualities, in plain logical contradistinction. But when the morphologist compares one animal with another, point by point or character by character, these are too often the mere outcome of artificial dissection and analysis. Rather is the living body one integral and indivisible whole, in which we cannot find, when we come to look for it, any strict dividing line even between the head and the body, the muscle and the tendon, the sinew and the bone. Characters which we have differentiated insist on integrating themselves again; and aspects of the organism are seen to be conjoined which only our mental analysis had put asunder. The co-ordinate diagram throws into relief the integral solidarity of the organism, and enables us to see how simple a certain kind of correlation is which had been apt to seem a subtle and complex thing."[9]

When we go so far as to assist integration not simply by promoting its natural predilection but by blending previously separate subjects, such as two different species, we are creating chimeras. Chimeras abound in ancient literature. To the student of mythology chimeras may be symbolic or even, occassionally, literal figures representing psychic energy on either the ethereal or the material plane. As though mythology were more than symbolic, referring to more than physiological quirk, and exceeding psychic manifestation, chimeras are being created in the laboratory. The question arises: Is the realm of mythology not only rematerializing but also becoming materialized not as a subject but as what that subject refers to, finding its path wherever it can, perhaps even through the unconscious workings of a particular science, genetics? Is the material world becoming mythologized? Genetics focuses on Source, here seen as the cell; any source may be considered whole unto itself. Genetic engineering divides Source as the World Parents are said to be divided. By virtue of its being manipulative it cannot be compared with natural cell division. It splinters the parts and rearranges them in the intellectual approach to source which characterizes breakdown on the mythological quest for accord with source and wholeness. In this sense the rearing of the head of mythology has permeated the sciences where, the sciences being largely of a materialistic orientation at present, mythology is expressed in terms of the variability of matter, which appears to be increasingly extensive. It should be noted that breakdown along the mythological quest can create an opening to wholeness which affords true accord.

More and more interesting dimensions appear in regard to the interrelationship between animals and humans. We are well aware of the antiquity of the employment of creatures to assist humankind. The horse drawing the plow, the mule or elephant carrying its passengers come perhaps most immediately to mind. The assisting function of these redefined aimals, beasts of burden, has been recently seen to subside in favor of mechanical devices such as the automobile and that machinery which is common to farming. Curiously, just as civilization approaches a seeming refinemcnt of mechanical devices and of its dependency on them, these mechanical devices are referred to with an anxiety concerned with their inorganic sterility, and civilization has beckoned through science once again to the potential assistance afforded by creatures other than ourselves or our machines. Hence, we have not only mutants such as drug-resistant bacteria but also more easily recognizable chimeras such as the bacterial "bug" devised by Ananda Chakrabarty, a hybrid created to help society clean up its oil spills through the sheer power of its appetite in conjunction with the preferred diet engineered into the nature of its functions.

The art of changing the disposition of an individual life form is well known especially in Celtic literature. It appears throughout the history of literature, including the "Histories" of Herodotus. Such a creature is actually referred to as a "changling" when the new disposition is an unpleasant or a supernatural one, especially, such as a beautiful baby's transformation into ugliness or an infant's sudden change from what had been seemingly permanent dumbness expressing itself in a lashing into a stream of exceptional conversation. Altered appetites, too, have been a symptom of the changling. Are we to consider Chakrabarty's oil-eater a changling? The ability to create a changling has been delegated to the likes of fairies and other supernatural beings, as well as to gods. Furthermore, the changling is not merely an affected individual but a substitute originating in the realm of the supernatural, in most cases, while the individual it replaces has been chosen to dwell among the less coarse divinities. Then might we consider the original coarse and potentially familiar oil-eater dies, if it exists anywhere naturally, so rapidly that it can not be detected because the gods can not be without it? Are we instead to assume that such a creature never did, in fact, exist in its sublimely coarse state because, perhaps, the gods required all of its potential, providing us, through the insistance of Chakrabarty, with a thing recognizably similar to what might have existed and which the gods possess but of an unnatural appearance therefore ugly or perhaps as full of potential surprises. Chimeras are not necessarily changlings, although engineered chimeras are not excempt from unforseeable behavior.

The appearance of live chimeras is moderate by comparison with many of those described by literature. Chimeras which are created by an experimentally conducted exchange between for instance, two species of mouse are no less suggestive than those of literature. The result of the research here referred to were interspecific chimeras. The experiment involved a chemically induced transference between the embryos of two different species of mouse. The work provides one solution to clonal analysis of cell lineages. In April of 1980 J. Rossant, Brock University, Ontario, and W. I. Frels, Jackson Laboratory, Maine, reported in volume 208 of *Science* this production of interspecific mammalian chimeras. It was the first report of entirely normal development of such chimeras. The chimeras were made by injecting the inner cell masses of the M. caroli species into M. musculus blastocysts. These blastocysts received the foreign cells without antagonistic selection against them. The work leads to several possibilities for advances in biological study. Among these is a cell marker system basic to studies in mammalian development.

One of the most bizarre forms of chimera is one mythology has never described. Illustrating a chimerically derived concept, electrical enzymes are the development (1980) of I. J. Higgins, Kent University, and Alan Hill, Oxford University. The biofuel cells, suggested as long ago as the turn of the century, are the result of methanol dehydrogenase extracted from bacterial cells as a tool whose task is to serve energy conversion. The original proposal suggestion that biofuel cells could provide energy conversion more efficiently than conventional generators of electricity. Methanol dehydrogenase is an enzyme that speeds the release of hydrogen. The enzyme adheres to the surface of solid electrodes where it catalyzes conversion of methanol to electricity.

Energy, its availability and appropriate functioning is the single greatest concern of the late 1900s. In every system mediating factors are required. The greatest mediating factor, humanity, provides the ultimate catalyst for the fused energies of the two great realms, the known and the unknown. The human mediator employs analysis and exploitation of intricately faceted life, a variable reality, to release energy.That analysis and exploitation have begun to turn inward for solutions obliterated in the external view which has been so extensively adventured and crowded already. Biology has turned to a deep interior, to a source of variability, for answers through the study of the cell.

Whether or not a product of the new biology is a chimera is doubtless overlooked by those hoping that genetic engineering will ease the strain on personal energy by providing drugs for diseases that reroute that energy from preferred outlets. Drugs which may be produced by genetic engineering are listed in Table 3–6. Manufacture of such drugs is part of what is called biotechnology. Biotechnology stands for the commercial component of the new science.

Stephen Turner, founder and president of Bethesda Research Laboratories, was asked to address the issues of business-related recombinant DNA technology for the New York Society of Security Analysts in June 1980. Asked about the growth potential for biotechnology, Turner responded with a prediction that the market for industrial applications of research in genetics would provide yearly increases of more than 50 percent between 1980 and 1990. These increases would occur in spite of the "volatility of technical leadership" which Turner classified as the first characteristic of biotechnology according to the business viewpoint. Venture capital pouring into the new field created a unique and continuous stream of biologically oriented articles in the *Wall Street Journal.* Scientists began leaving their academic homes for corporate atmospheres. The situation became so intense that it was felt that there were not enough genetic engineers to supply business with its surging demands. University-bound molecular biologists found dangling in front of them remarkable financial lures from new firms. The result was a threatening percentage of nonacademic researchers and pressure on universities to coordinate their research with the goals of the business world. Although most brokers shunned the long-sighted ventures, and stock value fluctuated, the draw on entrepreneurial and established business

Table 3-6. Diseases Amenable to Treatment by Drugs Produced by Genetic Engineering in the Pharmaceutical Industry.[a]

Disease or Condition	Drugs Potentially Produced by Genetically Engineered Organism
Diabetes[b]	Insulin
Atherosclerosis	Platelet-derived growth factor (PDGF)
Virus Diseases:	Interferon
Influenza	
Hepatitis	
Polio	
Herpes	
Common cold	
Cancer	Interferon*
Anovulation	Human chorionic gonandatropin
Dwarfism[b]	Human growth hormone
Pain	Enkephalins and endorphins
Wounds and burns	Human growth hormone
Inflammation, Rheumatic diseases[a]	Adrenocorticotropic hormone (ACTH)
Bone disorders, e.g., Paget's disease[b]	Calcitonin and parathyroid hormone
Nerve damage	Nerve growth factor (NGF)
Anemia, hemorrhage	Erythropoietin
Hemophilia[b]	Factor VIII and factor IX
Blood clots[b]	Urokinase
Shock	Serum albumin
Immune Disorders	Cytokines

[a] After: Office of Technology Assessment.
[b] Disease currently treated by the drugs listed.
*This remains conjectural.

instincts was sufficient to be described as effecting a whirlwind of activity.

Robert Reinhold wrote in the New York *Times,* June 22, 1980, that it was not a complete surprise for recombinant DNA technology perfected in university laboratories to be commercially exploited. The computer and electronics industries bridged university and technology just before World War II, and since then, technological innovation has continued to emerge from academic centers. But the bridge is no longer free from controversy. A wealth of research talent is being drawn by an expansive industry. Private biotechnological companies that have cropped up since 1960 are given in Table 3–7. Of the companies listed, the greatest quantity of employed Ph.D.s as of 1979 was a rising 50. In terms of creative brainpower for academic centers this figure represents an alarming percentage when the comparative loss for academic research is considered. Some researchers, however, find this to be a positive development. A shortage of university positions for Ph.D.s in biology began in 1970. Many of these researchers were working in less than desirable jobs at salaries beneath those of office secretaries simply because no other jobs were available.

The response of universities to biotechnology is not entirely enthusiastic. Apart from the exchange of brain power the relationship between university and industry has been aggravated by industry's eagerness for prestige and budget minimization. The award of the first U.S. patent in genetic engineering went to Stanford University with the consequence that commercial contenders in biotechnology were fired with the feeling that "the sedate world of university science may have walked off with the big prize. . . . Now, the go-go world of commercial biotechnology," wrote Marilyn Chase in the *Wall Street Journal,* must

Table 3–7. A Sampling of Private Companies with Biotechnology Programs.[a]

COMPANY	YEAR FOUNDED
Atlantic Antibodies	1973
Bens Bio Logicals	1979
Bethesda Research Laboratories	1976
Biogen	1978
Centocor	1979
Cetus	1972
Clonal Research	1979
Collaborative Genetics	1979
Collaborative Research	1961
Genentech	1976
Genex	1977
Hybritech	1978
Molecular Genetics	1979
Monoclonal Antibodies	1979
New England Biolabs	1974

[a] After the Office of Technology Assessment.

pay homage—if not substantial royalties. . . .'' Thus, independent advances have sometimes generated petty reactions, while collaborative efforts by universities and business are enacted as preferable alternatives.

The promises and realities inferred and actualized by studies in genetics affect many fields and industries. They are medically related and develop into new forms of microscopic surgery and new drugs for treatment. They are agriculturally, chemically, and environmentally relevant, and they become of daily importance when considered in terms of the food processing industry. In its applicability to agriculture genetic engineering is capable of revolutionizing important areas of the industry. It has been described as a key to future agricultural advances. The application of genetic engineering in agriculture may have a long-range potential rivaling the extent of its future in medicine.

Mendel's law of trait segregation provided breeders with the ability to predict the outcomes of hybridization. The laws also led to refinements in breeding methods. Breeding objectives were reached faster and with more precision. Production was significantly increased. For 80 years, since 1900, classical applied genetics has worked with the agricultural industry. It has led to an increase of yields; to an overcoming of natural breeding barriers; to an increase of genetic diversity for specific uses; to the expansion of geographical limits where crops can be grown; and to the improvement of plant quality.

Plant breeders have increased the productivity of many important crops used for food, feed, fiber, and pharmaceuticals. Their successful development of cultivars, cultivated varieties, to fit specific environments, and of production practices are responsible for this increase. Commercial development of hybrid corn in the early 1900s and of the ''green revolution'' wheats in the 1950s and 1960s are two examples of the effects of breeding on the supply of food for the world market.

Although there have been simultaneous improvements in farm management, pest control, and cropping techniques using herbicides, irrigation, and fertilizers, estimates indicate that applied genetics has accounted for as much as 50 percent of the harvest increases in the twentieth century. The primary objective of breeding has been, throughout history, the maintenance and improvement of crop yields. Other breeding objectives describe specific responses to the needs of local growers, to the demands of the consumer, and to the requirements of processing firms and marketing systems.

There are three major goals of crop breeding. They are frequently interrelated. The goals are as follows.

- Maintenance or increase of yields by selection of varieties in terms of their: pest/disease resistance; drought resistance; response to fertilizers; tolerance to adverse soil conditions.
- Increase of the value of yield by selecting varieties with such desirable traits as: increased oil content; improved storage qualities; improved milling and baking qualities; increased nutritional value, for instance, higher levels of proteins.
- Reduction of production costs by selection of varieties amenable to mechanical harvesting, which reduces labor requirements; varieties needing fewer chemical protectants or fertilizers; and varieties amenable to minimum tillage systems, which conserves fuel or labor, reducing the quantity of cultivation operations.

Commercialization of new varieties depends strongly on genetic variability. The variability affords selection and evaluation of specific traits. The six basic steps of a breeding system require selecting the crop, identifying the goal, choosing the method needed to achieve the goal, exchanging genetic material by breeding, evaluating the resulting strain in field conditions and correcting any perceivable deficiencies in terms of the breeding goal, and lastly, producing seed for distribution.

The biological characteristics of the breeding subject determine the plant breeder's approach. The breeder may choose one of two approaches, inbreeding or outbreeding, or use the two in combination. The breeder's choice is influenced by whether or not the plant naturally fertilizes itself or is fertilized by a neighboring plant. Repeated cycles of self-fertilization reduce heterozygosity. Numerous generations emerging from such a plant result in homozygosity, pure lines that breed true. Cross-fertilization, of course, creates a new mixture of genes. The combined approaches together produce a hybrid. Lines inbred for homozygosity are crossed to produce a parental line of "enhanced" genetic potential. The most vigorous hybrids are selected for further testing. The effects of hybrid vigor are expected to vary. They may include an earlier germination. They may also include increased growth rate or size. Greater crop uniformity is another effect.

The exchange or addition of genes through altering the number of a plant's chromosomes, or ploidy, is another method available to the plant breeder. When ploidy is increased plants usually gain full sets of new chromosomes. More than a third of domesticated species are polyploids. Such plants are generally larger than normal, their leaves broader, thicker, and their flowers, fruits, or seeds also larger.

Backcrossing can also create a variety superior in the eyes of the commercially oriented breeder. It is a process of lifting one or more desirable traits from an inferior subject. The breeder makes a series of crosses from the inferior to the superior plant while selecting for desired traits throughout successive generations. Self-fertilization of these plants results in a percentage of progeny that are homozygous for the genes being introduced from inferior parents and that are identical with the superior variety in all other respects. Single-gene resistance to pests and disease agents has been transferred successfully through this method.

The extent of genetic variation available for improvement by breeding is uncertain. Worldwide germplasm resources have been too heavily though not completely exploited. The ability of breeders to improve many already highly developed varieties has become more difficult. Height reduction in wheat, for instance, made huge contributions to its productivity but further improvement on this basis has been limited. Another crop, the potato, is derived mostly from survivors of the late blight epidemics of the 1840s in Europe and North America: "it seems likely that the genetic base was already somewhat narrow by the time modern potato breeding got under way. The five-fold increase in yield resulting from selection during the last 100 years of potato improvement has produced a group of varieties that are genetically similar and unlikely to respond to further selection for yield," states the National Research Council's Committee on Genetic Vulnerability of Major Crops (1972).[10] The level of homogeneity among various crops may make selection for higher yields more difficult in general.

New technologies used in combination with classical plant breeding techniques provide potentially useful tools. Technologies developed for classical plant breeding and for the new genetics are certainly not mutually exclusive. Both are tools for effectively manipulating genetic information through methods adapted from the recombination observed in nature. Indeed, it would seem odd if plant breeding were not one of the avenues for applicability of a technology that developed from experimental plant breeding.

Advances in genetic engineering permit plant breeders to bypass various natural breeding barriers that limit control in transferring genetic information. The new technologies do not necessarily offer agriculture the radical changes provided by recombinant DNA to microbiology. In theory, however, they will hasten the process of genetic refinement. They will perfect it.

New technological categories involve genetic transformations through cell fusion as well as the insertion or modification of genetic information through cloning of DNA. Most genetic transformations require enzymes to digest the impermeable cell wall of a plant. The process results in a cell without a wall, a protoplast. Protoplasts can fuse with each other and with other cell components. In theory, this ability permits an expanded exchange of genetic information.

These new technologies are approached in a three-part program. First, plant cells are isolated, established in tissue culture, kept alive. Second, genetic alterations are engineered in those cells; desired traits are selected if possible. Third, regeneration of the altered single cells is initiated, encouraging them to grow into entire plants. In microorganisms changes made on the cellular level are the goal. With crops, changes on the cellular level by themselves are virtually meaningless; they obtain meaning only when cells are grown into mature plants having new, desired characteristics.

Tissue culture involves cells grown from a plant in a supportive culture medium that keeps them viable, a process started with plant organs such as leaves or roots or with functioning aggregates of one type of cell such as epidermal cells in a leaf, or started with single cells. The process offers specific benefits to plant breeders. Tissue culture is basic to applications of new genetic technologies for plants in the same way that fermentation is crucial to microbial genetic technologies.

Tissue culture requires that isolated cells be induced to undergo repeated cell divisions in broth or gel. An amorphous cell clump, a callus, results. Readjustment of culture conditions when the callus appears allows the cells to undergo even further proliferation. As the resulting cells differentiate they grow, if all is successful, into well-organized tissues and organs, a complete plant normal in every regard except for the fact of its interaction with the engineer.

Production of 1000 plants from each gram of starting cells is not uncommon, for further subculturing of the callus allows mass propagation of a desired plant. One gram of starting carrot cells routinely brings forth 500 plants. The ultimate goal is easily imagined: Plantlets are placed in regular soil where they grow into fully functional mature plants appropriate for harvesting. The complete cycle, plant to cell to plant, produces plants on a far more massive scale and in far less time than conventional means permit. Commercial uses of tissue culture already exist.

Tissue culture can be used for long-term storage of specialized germ plasm. Storage involves the freezing of cells and types of shoots. Genetic material provided by culturing is stable. Reduced space requirements and decreased maintenance costs are also recognized as benefits of employing tissue culture. The cells remain viable for long periods this way. Carrot tissues frozen in liquid nitrogen, for instance, have been thawed after two years; and then regenerated into normal plants.

Plant cells in culture, similar to microorganisms in fermentation, can be engineered to work as "factories" to create plant products or by-products. Those commercially produced include camptothecin, which is an alkaloid with antitumor and antileukemic activity. They also include proteinase inhibitors such as heparin, and antiviral substances. In the future, other medicinals as well as flavorings and oils and even insecticides may be extracted from the cells. Some of the plants propagated through tissue culture, especially those relevant to agriculture and horticulture, are given in Table 3–8, while a list of various tissue culture programs is shown in Table 3–9. Clearly plant life is no longer limited to the caretaking of faeries and other supernaturals such as Sir Arthur Conan Doyle has described. But whether or not its relationship with humans truly brinks the "plant-people" described by the *National Enquirer* (July 1, 1980) in response to research attempting, unsuccessfully at that time, to create a hybrid from plant and human female cells is hardly debatable.

The genetic manipulation of cells in tissue culture and the selection of desired traits could extend botanical taxonomy to include new forms

Table 3-8. Some Plants Propagated Through Tissue Culture for Production or Breeding.[a]

AGRICULTURAL AND HORTICULTURAL:

Vegetable crops
- Asparagus
- Beet
- Brussels sprouts
- Cauliflower
- Eggplant
- Onion
- Spinach
- Sweet potato

Tomato

Fruit and Nut Trees
- Almond
- Apple
- Banana
- Coffee
- Date
- Grapefruit
- Lemon
- Olive
- Orange
- Peach
- Petunia
- Rose
- Orchid

Ferns
- Australian tree fern
- Boston Fern
- Rabbitsfoot fern
- Staghorn fern
- Sword fern

Bulbs
- Lily
- Daylily
- Easter lily
- Hyacinth

Fruits and Berries
- Blackberry
- Grape
- Pineapple
- Strawberry

Foliage
- Begonia
- Cryptanthus
- Dieffenbachia

Draceana
- Fiddleleaf
- Poinsettia
- Sivler vase
- Weeping Fig
- Rubber plant

Flowers
- African violet
- Anthruium
- Chrysanthemum
- Gerbera daisy
- Gloxinia

PHARMACEUTICAL:

ATROPA
Ginseng
Pyrethium

SILVICULTURAL (FORESTRY):

Douglas fir
Pine
Quaking aspen
Redwood
Rubber tree

[a]Source: Office of Technology Assessment.

Table 3-9. Representative List of Tissue Culture Programs in the United States.[a]

INDUSTRY	APPLICATION
Asparagus industry	Rapid multiplication of seed stock
Citrus industry	Virus elimination
Coffee industry	Disease resistance breeding
Pineapple industry	Mass propagation
Strawberry industry	Mass propagation
Chemical and pharmaceutical Industries	Biosynthesis of chemicals
	Propagation of medicinal plants
Land reclamation	Mass propagation
Ornamental horticulture	Mass propagation

[a] After the Office of Technology Assessment.

of plant life even if they are not cross breeds between plants and animals or humans. The insertion and deletion of the genetic information of plants is a possibility finer than the comparatively crude methods traditionally used in breeding. Approaches include the culturing of plant sex cells and embryos, protoplast fusion, and genetic transfer by DNA clones and foreign vectors. Screening for desired traits follows.

Culturing of plant sex cells, the egg and pollen, can increase efficiency in the creation of pure plant lines for breeding. Sex cells even in plants each contain only a single set of unpaired chromosomes. Plantlets derived from these maintain this single-set theme. Any genetic change therefore becomes apparent in the regenerated plant. A second paired gene would not ordinarily mask the effect of that change. Haploid plants have been produced in great quantity for more than 20 species. Simple treatment with the chemical colchicine can usually induce the duplication of their genomes and result in fully normal, diploid plants. Only one major crop, asparagus, has been bred by this technique. Overcoming the remaining technical barriers is equated with permitting use of the technique to enhance selection of trees and to create hybrids of important crops. Successful cultures of the sex cells of plants have been achieved for various important cultivars. The group included rice, wheat, barley, oats, sorghum, tomato, and tobacco. "However," as even the OTA is aware of a fact quite alarming to those mindful of the controversy in genetics, "because the technique can lead to bizarre unstable chromosomal arrangements, it has had few applications."

Embryos sometimes function as starting material in tissue culture systems. They are applicable to those systems requiring juvenile material. They hasten germination in species such as oil palms, which require up to two years to germinate in natural conditions.

Protoplast fusion, as already discussed, is a technique in which two entire protoplasts are united, or a single protoplast is joined to cell components from a different protoplast. The components are expected to fuse when they are mixed. The fulfilled expectation results in the formation of a single hybrid cell which may be induced to proliferate and to regenerate cell walls. Further culturing and regeneration into an entire plant is also possible. Protoplast fusion may be able to allow for an exchange of genetic information unrestricted by natural breeding barriers. Unfortunately for breeders, the method is limited by the instability of chromosome pairing.

Recombinant DNA technology provides the availability of selection and production of amplification of specific DNA segments. The "shotgun" approach requires the whole genome of the plant to be cut by one or more restriction enzymes. The DNA intended for transfer is attached to a plasmid or phage. It is carried by this vehicle into the plant cell. A gene coding for the protein zein, a major component of corn seeds, has, for instance, been spliced into plasmids and cloned in microorganisms. The hope is that the zein gene sequence is modifiable through this approach so that the nutritional quality of corn protein may be increased even before it is reintroduced into the corn plant.

Clones may perform an important role in genetically engineered forests. A model for the genetic engineering of forests is one in which the following are depicted: genetic material selected from a germplasm bank; insertion of the genes into protoplasts; regeneration of the cells from protoplasts; multiplication of cell clones. These are followed by mass production of embryos from cells; encapsulation to form seeds; field germination of the seeds; and finally, forests of new trees.

Technical constraints in genetic engineering encourage further and dedicated study for the full realization of potential benefits. Only limited success has, for instance, accompanied regeneration of whole plants from individual cells. The quantity of plant species that can be regenerated from tissue culture increased between 1975 and 1980, but those species include mostly vegetables, fruit and nut trees, flowers, and foliage crops. Wheat, oats, and barley as well as other important crops have yet to be regenerated.

Technical breakthroughs generally come on a species-by-species basis. Often they are not applicable to all plants. Even when the new technologies successfully transfer genetic information the result cannot be considered completely successful when the changes can be unstable. The observation of this instability has diminished the promises, temporarily, of protoplast fusion. The possibilities would be great, however, if whole chromosomes or chromosome fragments were able to be transferred in plants in which sexual hybridization has been impossible.

By November 1980, only one application seeking excemption from NIH guidelines and the release of corn treated by rDNA technology had been filed with the Office of Recombinant DNA Activities (ORDA). The extent to which possible ecological disruption from the release of genetically engineered plants, and attendant liability concerns, have prevented industry from directing further attention to commercial applications of the new technology. Progress in agricultural applicability, however, is not only present, it has grown far beyond expectations.

For the first time plant molecular biology was added to the list of meetings at the prestigious Gordon Conference in 1980. Other meetings have also begun to concentrate on it more. An influx of investigators from a variety of fields promises to enrich the different approaches used to solve the problems of the plant breeder.

Several areas will clearly feel the impact of genetic technologies given the appropriate and predicted breakthroughs. The batch culturing of plant cells in automated systems is already enhanced by the ability to engineer to select strains producing larger quantities of desired substances including drugs. In forestry the development of elite tree lines increasing yield is anticipated. Of potential in agriculture is the increased biological efficiency of many important crops, and plants not yet fully exploited may be improved more quickly.

Advances in reproductive biology have affected livestock production. Applied genetics in animal production has been a force behind the reshaping of traditional farms into modern production systems. Farms are increasingly specialized, capitalized, and integrated with off-farm services. Technologies of established usefulness include sperm storage, artificial insemination, estrus synchronization, superovulation, embryo recovery, embryo transfer and embryo storage, sex selection, and twinning. For anyone unfamiliar with reproduction beyond natural means these actually routine methods may seem quite uncommon. An example of their results is given in Table 3–10, which describes superovulation. The results of more experimental methods in reproduction include, for instance, the division in half of a two-cell embryo to obtain one pair of identical mouse twins. The speculative technologies for reproducing livestock animals include in vitro fertilization, parthenogenesis, cloning, cell fusion, recombinant DNA, and gene transfer.

The reproduction of farm animals is distinguished from that of single-cell organisms by two characteristics. Reproduction is sexual for the male, and male and female germ cells must generally be brought together. Also, reproduction is slower with animals because the generation interval is longer. The speed with which genetic improvements can be made in animals is limited.

The OTA states that every failure to capitalize on genetic potential is paid for by consumers. Genetic improvement can become a central objective throughout breeding programs. For this to happen, selection objectives must be identified and agreed on and conflicting interests must be brought together in such a way that a program serving all interests can be developed. Reliable, evaluative data on breeding stock is, furthermore, essential for breeders.

Improved germ plasm, a qualification equated with an increase in the nutritional content of fertilizer, improves the quality of the physical capital used on the farm. Further improvement of the germ plasm of all major farm animal species remains desirable for breeders.

Table 3–10. Superovulation Results in Farm Animals.[a]

Animal	Normally Expected	With Superovulation
Sheep	1.5	9–11
Pig	13	30
Horse	1	1

[a] After George Seidel, Animal Reproduction Laboratory, Colorado State University, Fort Collins, Colo.

The selection of desired characteristics causes specific qualitative changes. The efficiency of cellular information is perceived to be enhanced. Consideration for superior germ plasm during breeding decisions may upgrade a farmer's breeding or production stock.

Another area of applicability of the new technologies is aquaculture, the cultivation of freshwater and marine species, the latter also being known as mariculture. Aquaculture is roughly 6000 years old. Most commercial aquacultural operations depend on wild seed stocks. Reproductive control and genetic selection are therefore important. Selective breeding, hybridization, and the use of frozen sperm are employed. But aquaculture suffers from many problems, including insufficient species research.

The U.S. Department of Agriculture states that the goals of specific livestock research have significant potential for realization. These goals, both in the United States and in developing countries, include the control of reproduction and respiratory diseases, the development of animals that are considered genetically superior, the improvement of nutrition efficiency, and the increasing of reproductive performance.

The subtlety and extent of the change in breeding animals will result in their expression, if not always their actualization, of new qualities sufficient to allow them to be addressed as animals unfamiliar beside the physiological, genetic, behavioral maps of their predecessors. Within the context of the human diet this subtlety will doubtless go unperceived. Within the processes of agricultural production new problems will replace old ones, if that cyclic law continues to hold sway. Within the realm of physical assistors, however, the new technologies are justly of high regard.

In the food processing industry genetics can be used to design microorganisms that transform inedible biomass into food, either for human consumption or for animal feed. In addition, it can design organisms that assist food processing by acting directly on the food or by providing additives for food. The first food-related application of recombinant DNA was with bovine growth hormone (BGH). In March 1981 Genentech and Monsanto announced that in the previous year BGH had been produced by cloning the BGH gene from cattle into a microbe, an operation performed at Genentech for Monsanto. The Recombinant DNA Advisory Committee (RAC) had just approved the scaling-up of BGH at Genentech to quantities needed for feedlot studies. Interestingly, the synthetic bovine growth hormone diethylstilbestrol (DES) had been banned by the FDA only two years prior to the announcement. Natural BGH normally found in cattle is expressed by the microorganism employed by Genentech to produce it. Thus it is permissible in the eyes of the FDA and is expected not only to appear safe and effective in feedlot tests but also to result in more beef or milk per pound of feed.

A food ingredient category where yields could find improvement through the application of recombinant DNA technology is vitamins. Vitamins either are produced directly through fermentation, are chemically synthesized, or result from processes involving both fermentation and chemical synthesis. The most likely candidates for the application of genetic engineering are vitamins C, B_2, and B_{12}. Dr. Smolin of Hoffman-La Roche has stated, "Yields of vitamin C by fermentation are fairly low . . . it's conceivable that certain portions of the biosynthetic pathway, that is, those portions of the synthetic route to vitamin C which are biological, very well might be improved by genetic engineering technology . . . there are even indications that microorganisms can carry out the entire synthetic process."

On the horizon is also a whole new concept for upgrading yields in fermentation systems, especially the conversion of starch and cellulose to ethanol. This is made available by the genetic engineering of yeasts. Some organisms might be so genetically engineered as to obviate the necessity of preprocessing particular substrates. Yeasts such as *Saccharomyces cerevisiae* cannot hydrolyze starch without first being put through various processes. First of all, the starch must be broken down into glucose, a fermentable carbohydrate. Conceivably, an engineered organism could provide the desired starch-hydrolyzing enzyme as part of the cell's metabolism. Meanwhile, a continuous ethanol fermentation plant using segregated yeast strains was already placed under construction by National Distillers by 1981. *Bacillus subtilis* and *S. cerevisiae,* not the controversial *E. coli,* have been looked to as likely hosts for genes that may ultimately improve yields of food ingredients or in food processes. As *Food Engineering* quipped in their May 1981 issue, "Even though the organism (*E. coli*) is used in producing pharmaceuticals . . . food researchers are expected to opt for organisms generally recognized as safe."

Augmentation of the world's supply of protein has focused on microbial sources of protein for both animals and humans. Single-cell protein (SCP) is supplied in great quantities by bacterial and yeast cells, which are up to 72 percent protein-rich. The protein can be consumed directly or as an extraction processed into fibers or meatlike items. In terms of economics, genetic engineering had little impact on SCP processes until 1980. A variety of substances are now under consideration as raw materials for conversion. Among those considered are carbohydrates.

Single-cell protein can be obtained from a wide variety of microorganisms and raw materials. Of special note is that the nutritional value as well as the safety of each of the microorganisms varies widely. Single-cell protein, however, has the capacity to replace some traditional protein producers. Table 3–11 itemizes some of the raw materials already tested.

There are industries which would like to take full advantage of new technologies but are stymied for one reason or another. Hybridization and, more recently, cell fusion offer a variety of possibilities in the breeding of yeast used in baking, brewing, and wine making. Progress in these areas has, however, been limited by an insufficient quantity of suitable approved systems for use of recombinant DNA technology. Approval by the Recombinant DNA Advisory Committee, is necessary for extensive applied research in these industries. Other areas more readily meet approval.

Polysaccharides, polymeric sugars, are used by the food processing industry to both alter and control the physical properties of foods. Many polymeric sugars are derived from plant sources. Microbial

Table 3-11. Raw Materials Tested on a Laboratory or Small-Plant Scale.[a]

Agave juices	Peat (treated)
Barley straw	Pulp mill wastes
Cassava	Sawdust
Citrus wastes	Sunflower seed husks (treated)
Date carbohydrates	Wastes from chemical production of maleic anhydride
Meat-packing wastes	Waste polyethylene (treated)
Mesquite wood	

[a] After the Office of Technology Assessment.

polysaccharides have experienced limited use. They must offer new properties, meet all safety requirements, and also be readily available. The only one in large-scale commercial production has been xanthan gum. For many microbial polysaccharides an understanding of the biochemical pathway for synthesis and the systems controlling microbial production have not existed. Enzymes that may be employed for modification of polysaccharides have remained mysterious, the future of microbial polysaccharides pending. Meanwhile, however, recombinant DNA has been recognized as having the potential to make alternate hydrocolloids as opposed to the synthetic hydrocolloids such as modified celluloses which have been replacing costly and difficult-to-obtain gums.

In terms of coloring, carotenoids may possibly be made by fermentation. The next step would be to employ gene splicing to improve yields and concentrations. In another area, the potential for improving the fermentation of organic acids such as citric acid has gained attention. Production of citric acid is already very satisfying, but those excited by the concept of progress naturally perceive the possibility for even further improvement with open arms.

Enzymes, already derived from fermentation, themselves create improvements. Enzymes are produced for industrial, medical, and laboratory use. They are derived by fermentation processes that employ bacteria, molds, and yeasts by extraction from natural tissues. Fewer than 50 microbial enzymes have indicated their industrial importance. Patents, however, have been granted for over 1000. An increased interest in developing new enzyme products has met with some frustration. The discovery of a new enzyme is easier than the creation of an application for it that is also agreeable to biotechnology. Industrial enzymes are used in the detergent and in the food processing industries. They began to be used in quantity in the 1960s. Glucoamylase enzyme treatment began, then, to replace traditional acid treatment in the processing of starch. During the same period a stable protease enzyme was introduced into detergent preparations, helping to break down particular stains. By the 1970s glucose isomerase was converting glucose to fructose; it practically created the high-fructose corn syrup industry.

Larger quantities of existing enzymes and even new ones are promised by applications of biotechnology to fermentation processes. The role of genetic engineering in food processing is illustrated by the enzyme pullulanase. The enzyme degrades pullulan (a polysaccharide) to maltose or high-maltose syrups. The syrups give jams and jellies greater color and brilliance. They reduce heat discoloration in candies and prevent sandiness in ice cream by inhibiting sugar crystallization. The chief source of pullulanase is a bacterium, *Klebsiella aerogenes.* The bacterium is pathogenic. Efforts to apply genetics to improve the production and quality of pullulanase have not therefore, been significant.

The potentials, in summary, of genetic engineering to processes used in obtaining and enhancing food and feed are relevant to the following:

- Amino acids, enriching and flavoring agents such as monosodium glutamate
- Vitamins, enriching food
- Petrochemicals, food preservatives such as benzoic and propionic acids
- Short peptides, used as artificial sweeteners such as aspartame.
- Enzymes, involved in manufacturing processes
- Nucleotides, as flavoring agents
- Cellulose conversion, producing sugar and alcohol
- Polysaccharides
- Alternate hydrocolloids
- Colorants such as carotene derived from algae cultures
- Organic acids

Several conflicting forces impede an industry-wide impact of applied genetics on the food processing industry. Basic knowledge of the characteristics that could improve food is in but the early stages of development. The food processing industry is necessarily conservative in its research and in its development expenditures. Products are expected to satisfy FDA safety regulations. But the amount of required FDA testing may be reduced by transferring a desired gene into microorganisms already meeting FDA standards. Eagerness to employ recombinant DNA technology is quite evident in this concept.

"Genetic evolution is . . . an unconscious process. The interplay between organism and environment, however, is far more subtle than is indicated by this simple formulation. . . ."—René Dubos[11]

The possible role of the new biological technologies in the environment is another promising avenue for applied genetics. Mineral leaching and recovery is one of the areas in which genetic engineering can perform a helpful function.

Microorganisms generally interact with metals. Two interactions of industrial interest are the leaching of metals from ores, and the concentration of metals from wastes or dilute mixtures. Leaching allows extraction of metals from large quantities of low-grade ores, and concentration of metals allows methods for the recycling of precious metals as well as for the control of pollution caused by toxic metals.

The action of microbial leaching makes metals in ores soluble. It has long been known that dissolved metals can be recovered from mine and coal wastes. Operations based on this process date back to the eighteenth century. Bacterial leaching is used in large-scale operations in the United States in order to recover copper from wastes. Copper leaching accounts for over 10 percent of the total U.S. production.

Leaching, with which we are familiar in another sense in medicine, starts as water circulated through hundreds of tons of ore. Bacteria naturally associated with the rocks cause the metals to then be leached. The bacteria may act directly on ore or they may produce substances, such as ferric iron and sulfuric acid, which will extract the metal. Simply adding the acid is not as efficient as employing bacteria. Apart from the role of acid, direct bacterial attack on various ores appears crucial to the process. Some mineral leaching bacteria bind tenaciously to those minerals.

The leaching process in uranium mining presents the possibility of in situ mining. Conventional techniques to haul uranium ore to the surface are unnecessary. Microbial suspensions can extract the metal from its geological setting. Water is percolated through shafts where bacteria dissolve the metals, the solution is pumped to the surface, and the metal is "recovered." This "underground solution mining" came into use in Canada's uranium mines. It began there almost by chance. In 1960 the Stanrock Uranium Mine was found, after only two years of operation, to have large amounts of leached uranium contained in its underground water. Soon water circulation was part of the routine mining operation. Environmental benefits from this practice include a minimal disruption of land surface. The process is slower than the technology usually employed, however. The operation, though, is less costly because of the simplicity of this particular system. Agreeable bacteria replace the need for grinding machinery. The process makes possible readier mining of deeper and lower-grade deposits.

Bacterial leaching may also be used to extract sulfur-containing compounds from coal. The process results in coal of a lowered sulfur content. Sulfur-containing coals from certain areas are not as desirable, as other coals; they release sulfur dioxide during burning. A mixed population of different bacteria, as opposed to a single species, is most effective at removal of sulfur. The industry would therefore be pleased to see the genetic engineering of a sulfur-removing bacterium.

The leaching bacterium most scrutinized for its leaching properties has been *Thiobacillus ferroxidans*. This particular bacterium leaches copper. Other bacteria have also been identified. Very little is known about the precise genetic mechanisms operating the leaching activity in bacteria. A critical area requiring greater exploration is the chemistry of interactions between bacteria and rock surfaces. Further clarification of the genetic structure of the microorganisms is also necessary. The application of genetic technologies to mineral leaching remains speculative. Progress in research is slow because few laboratories have been active in it.

To develop engineered systems on a scale large enough to exploit their biological activities, microbial geneticists, geologists, chemists, and engineers need a constant interchange of ideas and information. This interaction is more readily achieved than a way of introducing large numbers of genetically engineered microorganisms into the environment which does not threaten ecological stability. There are, then, major obstacles extant for the use of genetic engineering in microbial leaching.

Microorganisms can concentrate metals from dilute solutions such as individual waste streams. The goal is to recover metals within a recycling process. Another part of the goal is elimination of any metal that may be a pollutant. Microorganisms bind the metals to their surfaces and concentrate them internally. Researchers at Oak Ridge Laboratory, Tennessee, have shown that microorganisms can be used for the removal of heavy metals from industrial effluents. Cobalt, nickel, silver and gold, uranium, and plutonium, as well as other metals, can be recovered. Of special relevance is application of the process in recovering metals from dilute solutions where nonbiological methods are not economical.

To consistently be reminded of the role of economics with regard to genetic engineering is to realize the very practical concerns related to techniques that may be required by industries whose survival is dependent on such considerations. How this affects genetic engineering itself will be part of the future industry shares with science.

The overall efficiency of the process of extraction may be increased by genetic engineering. An important step in this direction would be the selection of cells expressing the genetic ability to accumulate large amounts of specific metals.

Enhanced oil recovery (EOR) techniques are also methods that may be further enhanced by genetic engineering. These techniques employ chemical and physical methods of recovery. By increasing the mobility of oil, EOR makes the way easier for other forces to drive the oil out of the ground. Enhanced oil recovery is concerned mostly with oil in sandstone and limestone. Applied genetics may perform a major role in this area of EOR focus. Engineered microorganisms present possible recovery assistance.

The oil under consideration as a recipient of these processes is in reservoirs and pools at depths from only 100 feet to over 17,000 feet. It is absorbed on grains of rock and is usually accompanied by water and natural gas. Microorganisms can be selected or even genetically engineered to overcome many difficulties in the production of chemicals used in EOR.

Improved polysaccharides would have definite potential. Those obtained by fermentation would compete successfully with polysaccharides obtained from other sources such as seaweed. Marine pollution and weather changes would not disturb this method of production, which would be equally desirable because it could be geared to demand.

Microorganisms can produce substances applicable as cosurfactants in EOR. Already, micro-organisms producing surfactants have been identified. Genetically superior strains have been isolated. Chemicals, such as alcohols, that increase the rate of the formation and the stability of chemical/crude oil mixtures and agents that help to prevent precipitation of surfactants have also been produced by microbial systems.

Every oil field is unique. Each has its own set of characteristics. Salinity, pH, temperature, and porosity of the rock and of the crude oil vary. Any injected chemical would behave differently from one field to the next. It is difficult to ascertain sufficient knowledge of a well's characteristics to predict the nature of a chemical/crude oil interaction. The process of growing microorganisms in fermenters, extracting their chemical products, injecting the products into wells, may be bypassed, speculators have suggested, by injecting the microorganisms directly into the wells. The chemicals would then be produced in situ.

The geophysical and geochemical conditions in a reservoir rarely, however, encourage the growth of microorganisms. Microorganisms, like any other life form, whether it is used as a tool or not, require specific conditions in order to remain viable. They must be fed, and their environment, the levels of pH, temperature, oxygen, etc., must meet their needs at distances of hundreds of thousands of feet. The use of microorganisms by the oil industry has not always been encouraging. While the process for their use itself may be fine, its implimentation may be quite another matter and has sometimes led to discouraging experiences in the past. Such an experience occurred in the 1940s when injection into sulfite-reducing microorganisms with an inadvertently high-iron molasses as a carbon source frustrated the purpose with the formation of iron sulfide; the rock pores were clogged by the formation. Another case involved the development of yeast to break down petroleum. The size, however, of the yeast cells, 5–10 micrometers (μm), was enough to clog the 1-μm pores. Overall, greater precision was necessary. That precision is on its way. Geomicrobiology may have a viable future. Preliminary field tests have been encouraging.

The alteration (artificially induced or naturally occurring mutation) of microorganisms to produce chemicals for EOR has been successful. The alteration of those microorganisms that may be used in situ has been less successful. Applying recombinant DNA technology to either category would mean improvement in these endeavors.

The microbial production of oil muds or drill lubricants, and the treatment of oil after its recovery, are also due attention. In addition and in quite another avenue, pollution control, such as in the degradation and detoxification of human sewage, may be considered a target for applied genetics.

The concept of degradation as related not to sewage but to oil is one that has been explored by Ananda M. Chakrabarty as a concept suitable to the cross-barrier engineering of microorganisms. Chakrabarty is well known for the work he performed with four different strains of *Pseudomonas* to produce a chimera. Each of the strains bore characteristics describing their ability to degrade one of the four classes of chemicals found in oil spills. Important to the future of his work was that none of the strains was considered hazardous to human health, for the intent was their release, after engineering, into the environment. Chakrabarty engineered two of the organisms to be expert oil-eaters. His purpose was to create a single organism to particular specifications. Because of this kind of engineering the concept of "new life forms" splashed across news headlines and the Supreme Court was forced to consider whether or not life forms, new or not, are patentable.

Having discussed the potential of genetic engineering in the pharmaceutical and agricultural industries, in the food processing industry, and with regard to the environment, we are now able to continue to the chemical industry.

Three centuries before Christ, Alexandria began to replace Athens as the intellectual center of the ancient world. The Egyptians had long been practicing the art of preparing cheap alloys resembling silver and gold. Aristotle's influence gradually developed the idea that base metals such as lead could be changed into gold, an idea quite apart from simple resemblances among metals. Experimentation began in search of the transforming essence. The essence was never discovered but some experiments did actually succeed in preparing materials very closely resembling at least the superficial characteristics of gold.

In the seventh century A.D. the Arabs conquered Egypt. Alchemy became an Arabian art. An outstanding Arabian alchemist, Geber (721–813) believed all metals to be composed of sulfur and mercury. His influence was felt as late as A.D. 1500.

Paracelsus stated that three elements, earth, air, and water, were represented by salt, sulfur, and mercury, respectively. Paracelsus was daring in many ways and was the first of the iatrochemists, chemists who devoted their lives to the search for medicines. But from his time onward alchemists came to be thought of as rogues.

The first distinction between mixtures, compounds, and elements, as is taught to every school child, was made by Robert Boyle (1627–

1691). Boyle was an early member of the Invisible College, which adopted the experimental and inductive method advocated by Francis Bacon. The group was similar to others springing up almost simultaneously in Germany, France, and in Italy. The Invisible College developed into the highly regarded Royal Society.

In 1661, Boyle, who also introduced the term *analysis,* published the volume, *The Sceptical Chymist,* in which his famous distinction between mixtures, compounds, and elements was given as follows:

> "I mean by elements certain primitive and simple, or perfectly unmingled bodies; which not being made of any other bodies, or of one another, are the ingredients of which all those called perfectly mixt bodies are immediately compounded, and into which they are ultimately resolved."

Because of negative attitudes to alchemistry, and especially because of the still youthful nature of chemical knowledge, chemistry was a long time in becoming a respected science. Benjamin Thompson (1753–1814), known as Count von Rumford, has been credited with having placed scientific research in general at the service of the community. In 1799 he collaborated to found the Royal Institution of Great Britain, a society dedicated to the advancement of physics and chemistry. Professorships were endowed to ensure that scientists would have ample means at their disposal. The researchers were not to be burdened with teaching duties but were expected to lecture about their own work and about scientific progress in their field. The foundation of the society was a turning point in the history of science. Chemistry and physics were no longer subordinated to natural philosophy. They had become full-fledged sciences.

The chemical industry has become one of the largest and most important industries in the world. The basic function of the industry is the transformation of low-cost raw materials into end-use products considered to be of comparatively greater value. Petroleum, coal, minerals (phosphate, carbonate), and air itself (oxygen, nitrogen) are the most important raw materials. The greater part of the industry is involved in the production of inorganic chemicals such as lime, salt, ammonia, carbon dioxide, chlorine gas, and hydrochloric and other acids.

Applied genetics may play a major role in the chemical industry by conversion of biomass and carbohydrates from natural sources into various chemicals. Biology thus performs a dual role: provision of both raw materials and a process for production. Approximately one third of the industry produces organic chemicals. These include plastics, synthetic fibers, and synthetic rubber. It is this portion of the industry that is most under scrutiny for the possible applications of biotechnology. .

Applied genetics has contributed to the economic viability of the process of fermentation which still produces a few chemical entities. Production of citric and lactic acids and various amino acids has also benefited from genetics. Between 1960 and 1980 manufacture by fermentation was reduced by competition from chemical synthetic processes. Lactic acid is produced both synthetically and by fermentation.

Of all of the amino acids 20 are incorporated into proteins manufactured in cells; others serve specialized structural roles and are important as metabolic intermediates or neurotransmitters. All of the amino acids are used in research, in nutritional preparations, and in the preparation of pharmaceuticals. In a specific instance, glutamic acid manufactures monosodium glutamate, an ingredient in oriental cooking. There is for monosodium glutamate some, if not great, potential for the use of applied genetics to improve its yield. Table 3-12 describes the common sources for amino acids and the potential here for biotechnology. Microbial production of all the amino acids has been studied; glutamic acid and *L*-lysine are, however, produced in the most significant quantities. They are produced by fermentation processes.

Fermentation and enzyme technologies promise to create an impact on chemical process development. The transition from nonrenewable to renewable raw materials will be affected by fermentation. Technology involving enzymes will promote the entrance of fermentation-derived products into the chemical conversion chains, competing directly with traditional chemical transformations. Fermentation replaces various production steps. It could become a technology that is complementary to the overall manufacture of a chemical.

The efficiency of biological processes eliminates or simplifies many conventional waste, disposal, and product-recovery problems. By their very nature, biologically based chemical processes create some biodegradable waste products, products themselves valuable as sources of nutrients. Genetic engineering is expected to have an impact on three specific processes:

Table 3-12. Commercially Produced Amino Acids.[a]

AMINO ACID	COMMON SOURCE	POTENTIAL FOR BIOTECHNOLOGY
Alanine	Hydrolysis of protein; chemical synthesis	—
Arginine	Gelatin hydrolysis	Fermentation in Japan[b]
Asparagine	Extraction	—
Aspartic Acid	Bioconversion of fumeric acid	Bioconversion
Citrulline	—	Fermentation in Japan
Cysteine	Extraction	—
Cystine	Extraction	—
Dihydrophenylalanine (DOPA)	Chemical	—
Glutamic acid	Fermentation	De novo: *Micrococcus glutamicus*
Glutamine	Extraction	Fermentation in Japan
Histidine	—	Fermentation in Japan
Hydroxyproline	Extraction from collagen	—
Isoleucine	Extraction	—
Leucine	—	Fermentation in Japan
Lysine	Fermentation (80%) Chemical (20%)	De novo: (fermentation) *Corynebacterium flavum*
Methionine	Chemical from acro-acrolein	—
Ornithine	—	Fermentation in Japan
Phenylalanine	Chemical from benzaldehyde	Fermentation in Japan
Proline	Hydrolysis of gelatin	Fermentation in Japan
Serine	—	Bioconversion in Japan
Threonine	—	Fermentation in Japan
Tryptophan	Chemical from indole	—
Tyrosine	Extraction	—
Valine		Fermentation in Japan

[a] After the Massachusetts Institute of Technology.
[b] Japan remarkably predominates this picture. It is a nation determined to make rapid progress in genetic research. The frantic public/private coordination characterizing Japanese technological advances in general is present in its biotechnology as well. Japan has determined to sponsor the first major international conferences on DNA research (1981).

- Aerobic fermentation. This can produce specialty chemicals such as enzymes, vitamins, pesticides, growth regulators, and amino and nucleic acids.
- Anaerobic fermentation. The products of this process can include organic acids, methane, and solvents.

- Chemical modification of products from these two processes of fermentation.

Antibiotics, growth factors, and enzymes as well as other complex biochemicals were already being made by aerobic fermentation by 1981. Meanwhile, 40 percent of the ethanol manufactured in the United States was already being produced with anaerobic fermentation. Chemical production technologies using high temperatures and pressures might be entirely replaced by biological technologies which operate at atmospheric pressure and ambient temperature. Examples of the ways fermentation has expanded into the chemical industry arc given in Table 3–13.

The overall distribution of applied genetics in the food, chemical, pharmaceutical, and specialty chemical (food ingredients, reagents, enzymes) industries indicates that the chemical industry receives the greatest part of the total, 17 percent, before the pharmaceutical industry. Genetic engineering will develop new techniques for synthesizing many substances. No industry is likely to be directly displaced, but instead may be enhanced by what the Office of Technology Assessment refers to with something of an understatement as "simply another industrial tool."

Table 3-14. Expansion of Fermentation into the Chemical Industry.[a]

METHOD	EXAMPLES
Areobic Fermentation	
Enzymes	Amylases, proteases
Vitamins	Riboflavin B_{12}
Pesticides	*Bacillus thuringiensis*
Growth regulators	Gibberellin
Amino acids	Glutamic, lysine
Nucleic acids	
Other acids	Malic acid, citric acid
Anaerobic Fermentation	
Solvents	Ethanol, acetone, *n*-butanol
Acids	Acetic, propionic, acrylic

[a] After the Office of Technology Assessment.

4
Controversy

"Our trouble is the object Greek science we use . . . it has cut itself off from an adequate understanding of the Subject of Cognizance, of the mind."—Schrödinger

"Being is eternal; for laws there are to conserve the treasures of life on which the Universe draws for beauty."—Goethe

Given the phantasmagoric ploddings of the twentieth century, one is tempted to hope that there are laws to conserve the boundaries of familiar life structures. We seek to avoid any encumbrance more bizarre or unpleasant than those already experienced, and yet we seek also to retain our physiological status, we are ill prepared to look to a future of either bodiless brains or disembodied souls.

An effort to create laws to conserve laws occurred in biology during the early 1970s when scientists themselves declared a moratorium on genetic engineering research. Deep concern had already begun to spread by the time Maxine Singer, cochairperson of the Gordon Conference on Nucleic Acids (New Hampton, New Hampshire, June 1973), at a special session concluding the conference, sharply etched as follows a portrait of the two emerging faces of DNA research.

"We all share the excitement and enthusiasm of yesterday morning's speaker, who pointed out that the scientific developments reported then would permit interesting experiments involving the linking together of a variety of DNA molecules. The cause of the excitement is two-fold. First, there is our fascination with an evolving understanding of these amazing molecules and their biological action. Second, there is the idea that such manipulations may lead to useful tools for alleviation of human health problems. Nevertheless, we are all aware that such experiments raise moral and ethical issues because of the potential hazards such molecules may engender. . . . Because we are doing these experiments, and because we recognize the potential difficulties, we have a responsibility to concern ourselves with the safety of our co-workers and laboratory personnel, as well as with the safety of the public. We are asked this morning to consider this responsibility."

At this time a very alarming situation became apparent. There were no guidelines, under any kind of directive, for the research being carried out. The research took place wherever and whenever, however it could, even making its way into a rundown Volkswagen bus converted into a primitive laboratory parked on a side street in Washington, D.C. Within a few years Harvard was in an uproar. The dilemma was soon swept into the heights of ethics, out of which one of its debates emerged on the platform of a local Cambridge, Massachusetts church. On a simultaneously more and less serious scale, DNA splicing kits (their contents nonliving) were made available through mail-order coupons in magazines responsive adolescents may easily be imagined as devouring. The danger, presumably, of such splicing kits can hardly be greater than that of "DNA" and "RNA" pills sold as over-the-counter vitamins and guidebooks on DNA diets which were prevalent in the first years of the following decade. These are the hoax responses to real potentials whose importance and safety they complicate with extended anxiety. Among the issues of real concern have been:

1. *The potential hazards of certain possible combinations of DNA and the vehicles used for their propagation:* Stanley Cohen early recognized this problem. He asked colleagues to avoid making combinations with tumor viruses or any other DNA that would make the *E. coli* bacteria propagating them too resistant to antibiotics. He also asked that plasmids not be passed around without his knowledge, so sure was he of the potential hazards. When this self-regulation proved insufficient, Cohen, as had Paul Berg, became truly convinced of the danger.

2. *Technique:* Systems of ventilation, pipetting, and other standard laboratory equipment and procedures were seen as possible means of accidentally infecting researchers and the public with antagonistic microorganisms possibly made more antagonistic by intervention in their natural processes. Since the addition or deletion of DNA segments in agents of disease was performed in an effort to determine the role of various segments in the activity of the agent, an agent of increased risk as well as one of diminished risk might conceivably result during the course of an experimental study.
3. *Researcher Responsibility:* The responsibility of researchers was seen as being dependent on either enforcement of regulations or self-regulation of research with wider recognition of dangers. Not untypically, both soon prevailed.
4. *Ethics:*
    A. *Cloning,* especially as applied to humans, became a frightening if still distant (in terms of humans) possibility recognized mostly by the public willing to look so far ahead.
    B. *New life forms,* as they have been called, brought a thoroughly unexpected and problematical dimension to the controversy.
    C. The potential of *eugenics* simply reemerged in a more elaborate form.
    D. *Corporate participation* was seen as a threat to the puristic nature of research: truth-seeking.
    E. Out of all this, *manipulation* itself, as well as the quality of *sympathy* came under fire.

In the morning session following the June 1973 Gordon Conference a letter was drafted to the presidents of the National Academy of Sciences and the Institute of Medicine; it asked that an ad hoc group be constituted to study the potential risks involved in experimental research. *Science** soon published the letter, and an even wider spectrum of the scientific community became aware of the growing concern. The letter read:

* *Science*, Vol. 181, 1973, p. 1114

"We are writing to you, on behalf of a number of scientists, to communicate a matter of deep concern. Several of the scientific reports presented at this year's Gordon Research Conference on Nucleic Acids indicated that we presently have the technical ability to join together, covalently, DNA molecules from diverse sources. Scientific developments over the past two years make it both reasonable and convenient to generate overlapping sequence homologies at the termini of different DNA molecules. The sequence homologies can then be used to combine the molecules by Watson-Crick hydrogen bonding. Application of existing methods permits subsequent covalent linkage of such molecules. This technique could be used, for example, to combine DNA from animal viruses with bacterial DNA, or DNA's of different viral origin might be so joined. In this way new kinds of hybrid plasmids or viruses, with biological activity of unpredictable nature, may eventually be created. These experiments offer exciting and interesting potential both for advancing knowledge of fundamental biological processes and for alleviation of human health problems.

"Certain such hybrid molecules may prove hazardous to laboratory workers and to the public. Although no hazard has yet been established, prudence suggests that the potential hazard be seriously considered.

"A majority of those attending the Conference voted to communicate their concern in this matter to you and to the President of the Institute of Medicine (to whom this letter is also being sent). The conferees suggested that the Academies establish a study committee to consider this problem and to recommend specific actions or guidelines, should that seem appropriate. Related problems such as the risks involved in current largescale preparation of animal viruses might also be considered.

"Maxine Singer, National Institutes of Health
Dieter Söll, Department of Molecular Biophysics and
Biochemistry, Yale University."

The National Academy asked Paul Berg, Stanford University, to chair the suggested committee. Regulation of genetic engineering had become a topic worthy of formal consideration. The two most significant initial results of the Berg committee were, first, the recommendation that an international conference be convened; accordingly, the committee drafted a letter to *Science* asking for international self-regulation; furthermore, the committee wrote to request establishment of an advisory committee to consider the vast specifics of the rapidly developing situation. The advisory committee thus formed resulted in the production of volume upon volume of discussion of the regulations, which were ever under revision; the material was made available for monitoring by both professionals and the public.

The path to the historically dramatic Asilomar Conference had been laid. On the precipice new concerns emerged. Among these addenda of potential hazards was the production of new disease, whether that production be inadvertent or deliberate. In his *Double Image of the Double Helix,* Grobstein likens this to the 1930s, when nuclear physicists foresaw the atomic bomb. Referring to myth in its more recent definition as falsehood, Grobstein writes that, "The two experiences make clear that knowledge is as close to intervention as imagination is to myth—whether for good or evil. The seeker of 'truth' singlemindedly concentrates on expanding knowledge. He or she usually neither sees, speaks, nor plots good or evil. But new objective truth, when refracted in a multifaceted social structure, can emerge with conflicting shapes and implications." The initial concern over genetic engineering expressed a desire for prohitions, not just modifications.

"Anyone affirming immediate disaster is a charlatan, but anyone denying the possibility of its occurring is an even greater one."—Erwin Chargaff

The initial outcome of the Asilomar Conference, before any answer had been received to solicitous letters such as that to the National Institutes of Health (NIH) for a risk assessment committee, was a total moratorium called on certain experiments believed to be of a dangerous or even potentially dangerous nature. Some of these are given in Table 4–1, which reflects a portion of what became the first guidelines for research employing the techniques of recombinant DNA. The moratorium was called in July 1974. Six years later recombinant DNA was firmly established as a powerful technology with a growing number of researchers employing it and with rising industrial interest in its potential.

The foundational origins of unrest about the research in recombinant DNA stretch as far back as 1971. In June of that year a workshop in tumor viruses was held at Cold Spring Harbor Laboratory. Workshop attendance included Janet Mertz, a graduate student working in Paul Berg's laboratory at the Stanford University Medical Center. In a conversation with Robert Pollack of the Cold Spring Harbor Laboratory staff Mertz described one of Berg's upcoming ex-

Table 4-1. Various Recombinant DNA Experiments and Required Containment as Specified by the Initial Guidelines.[a]

DNA INTO *E. coli*	CONTAINMENT (INCREASING INTENSITY UPWARD FROM P[1] LABORATORY CONDITIONS; P REFERS TO PHYSICAL CONTAINMENT, EK REFERS TO THE LEVEL OF DISARMING MODIFICATION FOR E. COLI STRAIN K[12])
Nonembryonic primate tissue	P3 EK3 or P4 EK2
Embryonic tissues or germ line cells from primated	P3 EK2
Prokaryotes that exchange genes with *E. coli*—	
Nonpathogens	P1 EK1
Low-risk pathogens	P2 EK1
Moderate-risk pathogens	P2 EK2
Prokaryotes that *do not* exchange genes with *E. coli*—	
Moderate-risk pathogens	P2 EK2
Animal virus	P4 EK2 or P3 EK3
Plant virus	P3 EK1 or P2 EK2
DNA from highly pathogenic organisms and oncogenic organisms of moderate risk	BANNED
DNA of plant pathogens which may increase virulence or host range; DNA containing genes for toxins; DNA transferring drug resistance to organisms not naturally acquiring it: ALL	BANNED

[a] After a report prepared for the U.S. House of Representatives.

periments. The experiment was designed to recombine Simian-Virus-40 (SV 40) with phage lambda DNA. The hybrid would be inserted into *E. coli*. Pollack was concerned about the potential of such a mixture to cause tumors in humans. He telephoned Berg to discuss the possible danger. Berg did not immediately perceive his work as dangerous. But, by the early part of the following year, 1972, concern over his experiments had spread. Growing wary and unable to assign zero risk to his experiments, however novel such an assignment would be, Berg decided to postpone his plans for the experiment. Subsequently, he asked Pollack to organize a meeting for the purpose of discussing the risks of tumor virus work. During the meeting, the Conference on the Safety and Hazards of Tumor Viruses, January 22, 1973, Berg requested that the NIH assist with the concerns. He asked that epidemiological experiments be conducted to determine the risk of oncogenic viruses to individuals. Berg's request was actually quite limited by comparison with the scope the issues eventually assumed. His request, however, went unheeded.

Unrest continued. Within six months of the January conference on tumor viruses another important conference was convened. It was the Gordon Conference on Nucleic Acids at which conference cochairpersons Maxine Singer, National Institute for Allergic and Infectious Diseases, and Dieter Söll of Yale University received the suggestion for a short discussion of potential risks of recombinant DNA experimentation. It was this discussion which led to the formulation of the letter that was sent to the presidents of the National Academy of Sciences (NAS), the Institute of Medicine, and to *Science,* the publication of the American Association for the Advancement of Science (AAAS). The ensuing moratorium led to an international conference held at Asilomar and sponsored by NAS. One hundred and fifty-five scientists, lawyers, and media representatives attended the four-day conference which opened February 24, 1975. The five topics covered at the Conference were:

1. Ecology of plasmids and enteric bacteria.
2. Molecular biology of prokaryote plasmids and their use for molecular cloning.
3. Synthetic recombinants involving animal virus DNAs.
4. Synthetic recombinants involving eukaryotic DNAs.
5. Ethical and legal concerns.

Research guidelines were drawn up. They included four risk categories ranging from minimal to high. Conferees voted to ban experiments involving the DNA of pathogenic organisms and to extend the ban until suitable vectors became apparent or were developed. The growing concern had made self-regulation a more probable and widely acknowledged responsibility than the history of science had ever known.

Before the concern over potentially hazardous experimentation could explode into a controversy over the freedom of research, the NIH had held four meetings to develop its own guidelines. The first meeting, chaired by NIH Deputy Director DeWitt Stetten, Jr., simply concluded to employ the interim guidelines established at the Asilomar Conference. This deserves more appreciation than it might engender if excluded from prior data. The NIH now heeded the concern that had been growing since 1971. The risks of genetic engineering were getting active attention from a broadening circle of professionals. At the second meeting of the NIH, May 1975, a subcommittee chairman, David Hogness, Stanford University, accepted the task of developing more specific guidelines. The "Hogness" guidelines were revised at a third meeting held in Woods Hole, Massachusetts. This occurred just two months later. Momentum had been gained. At Woods Hole a great deal of conflict arose over determining whether the guidelines were too lax or too stringent. The conflict was so great that Chairman Stetten was given to search out a clearer perspective by circulating the guidelines within the scientific community. Response suggested the inevitable, and in December 1975 the fourth NIH meeting convened at La Jolla, California. By this time Elizabeth Kutter, Evergreen State College, Washington, had accepted the task of seeing that a newly revised set of guidelines be prepared. At the La Jolla meeting the "Hogness," the "Woods Hole," and the more recent "Kutter" guidelines were reviewed. Formulation of a final set of guidelines was begun.

A Washington, D.C. meeting of the NIH in February 1976 discussed the guidelines. At last, agreement permitted publication of guidelines for recombinant DNA research in the *Federal Register* of June 7, 1976. For anyone unfamiliar with the subject, it should be noted that it is a misconception to think that publication of the guidelines indicated their finality. The vast quantity of revised literature available from the NIH on this subject indicates the nature of the guidelines. They are in a continuous state of flux due to further research, response to research, and response to the guidelines themselves.

"If sufficient local banning or paring down of recombinant work occurs, researchers may be driven underground like the bootleggers of the Prohibition era."—*Wall Street Journal**

The banning and limiting of experiments has frequently been felt, especially by those outside of the groups intrumental in bringing them about, to encumber rather than to protect researchers; although even some of those initially involved in the concern, such as James Watson, later relaxed their views as well. On a local scale the total banning of

* Bronson, Gail, "Controlling Gene Transplants," *Wall St. Journal,* May 11, 1977.

recombinant DNA research has become an arbitrary issue. By the first few months of 1981 many cities and towns, fearing harmful leakage of bacteria, became reluctant to allow DNA research within their boundaries. Three towns in the United States, including Waltham, Massachusetts, worked independently toward local banning of all such research.

The fluctuating stringency of the guidelines has meanwhile continued since they were first proposed. From month to month and year to year a clear statement has been frustrated, as indicated by headlines such as those appearing over articles in *Nature* and other professional publications, with shifts from "NIH director unlikely to grant exemption from controls for DNA experiments," to "Committee Splits over Downgrading of E. coli containment." The shifts stem from within the scientific community, especially. The effect of the public on genetic engineering is comparatively small in spite of various organizations representing public concerns or of the reluctance of various towns to encourage the research. The willingness of the NIH to include the public voice in its redrafting of the guidelines is easily perceived as a token gesture considering the specialized nature of the subject, yet can also provide an example of public negligence considering the limited response to widely available, if quietly identified notices, such as small floundering articles in the *New York Times,* inviting public participation in guideline revisions.

"DNA-Test Curbs May Be Eased
. . .
The public has 60 days to comment
before the guidelines take effect."—UPI July 28, 1978

The public has been represented to the NIH by various organizations. These include the Coalition for Responsible Genetic Engineering, New York, and the Center for Study of Genetics and Ethics sponsored by the Peoples-Business Commission of Washington, D.C. Public concern and representation in the issues of genetic research, if it is not direct involvement, ranks, in terms of scientific issues, with the issues of ecology and nuclear power. It has begun attracting proponents and dissenters. Because the issue is yet a delicate and complicated one, rallies are not so easily formed over genetic engineering as over, for instance, nuclear power. The burden rests with concerned researchers, ethicists, and organizations representing the general public.

In 1977 when "efforts to control possibly hazardous genetic research are foundering," the *New York Times* reported in an editorial that, "The slowdown of the Congressional push toward regulation will allow time to devise a sound approach. There is no need for haste. The vast majority of researchers are voluntarily following the guidelines already, and virtually all experts agree that the chances of catastrophe are remote. But legislature is needed to make the guidelines binding on researchers not funded by the Federal Government and to establish better enforcement machinery. Regulation is too important to be left to the scientists."[1]

The public's fear seemed to be that the excitement of discovery and the potential for discovery in this field might sway scientific thinking into a dangerous liberality. This fear was responded to in a number of ways. "If we wish to maximize beneficial research, without jeopardizing either public welfare or public concern," wrote Bernard D. Davis, Harvard Medical School, "we must find ways to work out more clearly the role of the community of experts in ascertaining the facts as objectively as possible, and the subsequent role of the broader public in making value judgments and policy decision."[2] What Davis suggests is "avoiding premature discussion of unproved possibilities," a tactic the medical profession has been known to employ. Such a tactic underestimates what the goal of individuals and society might hope to attain. The possible as well as the probable are necessary to any intelligent analysis. Analysis, based on as much objectivity and long-range foresight as possible, has continued.

Laws to govern laws, or more precisely, guidelines to govern scientists, have indeed been developed (1976) by the National Institutes of Health. It cannot be said that the pleas for regulation continued to fall on deaf ears, for but a few years was needed for them to be heard and appreciated. The guidelines themselves, however, have caused a controversial stir.

Even as they began to take shape in the wake of the moratorium and research could continue, the question was, as *Science* put it, "how could the public be informed the moratorium was at an end when the unknown hazards that caused it to be invoked in the first place were just as unknown as before?"[3] Several years after the moratorium had been lifted and in the midst of the growing laxity of regulations, with short-term research masking potential hazards with an acceptable success at safety, and as competing scientists burning with curiosity demanded the greatest possible freedom in which to experiment, the rules began to be broken. A highly publicized incident of lawbreaking involved work in thalassemia performed by Martin J. Cline, University of California. The first attempt at human gene transplantation (July 1980) since the prohibitory stipulations of the NIH guidelines had gone into effect was the unauthorized application of recombinant DNA technology to two young women in Jerusalem and Naples. The attempt seems at once rash and courageous. There was a time, before antibiotics, when in vivo testing was popular. In vivo testing of microbial preparations was accepted and common practice throughout the history of medicine and until only very recently. Times have, thankfully, changed. Technology has changed. Attitudes have changed. The caution associated with a technology as intimate as genetic engineering is not unjustly founded.

Cline's human experimentation in gene transplantation was rash because insufficient experimentation on animals preceded it. This fact led UCLA to deny permission to Cline to carry out the work on humans. That Cline removed himself to areas not bound by this reserve, Israel and Italy, is an act of bravado shouting, "The sick must be healed!" Whether or not the sick have been healed by Cline's operation remains to be seen, as does the possibility that inadvertent harm might have instead resulted from what was less valid microsurgery, employing techniques used experimentally with mice, than it was serious insult to sober research. As a result, Cline became the prototype lawbreaker among genetic engineers. He was reprimanded by a UCLA and NIH very much aware of the anxiety over genetic engineering, and finally left his post at UCLA. Cline's fellow researchers were not pleased. They termed his research premature and dangerous for application to humans. Cline's performance occurred in 1980. It could not defeat or threaten, though perhaps it will encourage, the still open ended history of concern.

Toward the end of 1977, until the patenting of life reinspired anxiety, 15 years of erupting concern began to subside. The regulation of research contributed to the relative calm. The implementation of guidelines evidenced both their applicability and the need for their constant revision. An elaboration of those concerns and others since them follows.

Molecular biology was long considered "the practice of biochemistry without a license."[4] More recently, molecular biology has been considered the practice of playing God, also without license. It has pried into the most closely held secrets of biological life. Socrates is said to have repudiated Anaxagoras for thinking that prying into nature's secrets was forbidden by the gods. In the early 1900s, R. C. Punnett described an attitude he shared with William Bateson, that "any tampering with [chromosomes] by way of breakage and recombination was forbidden."* The attitude shared by Punnett and Bateson, based on a technical misunderstanding, is today shared for ethical reasons by many people in and outside of science.

Robert Sinsheimer, Caltech's biology chairman, has stated that "biologists have become, without wanting it, the custodians of great and terrible power. It is idle to pretend otherwise." After a decade of research the question of whether or not recombination research could promote existing forms of disease or unleash new ones has become hampered by a problem similar to that afflicting foes of nuclear power: Either the hazard has not happened yet or it has not made its presence felt as of yet. This applies not only in terms of disease possibly arising from rDNA research but also from all the many other concerns, including damage to the global gene pool. Is there an invisible new disease demon lurking in the confines of a time-lag of which we are unaware? The question of "will there be," is even more difficult than is the prediction of the course of a bioengineered organism whose individual path is no more, if not less, foreseeable than that of a "natural" organism. But the questions, in the midst of existing and burgeoning discovery, have lost the force they had nearly 10 years ago, in the 1970s, when the moratorium was replaced by guidelines.

As the new technology and its guidelines emerged into the limelight unpredictability in research results became a complex issue beside guidelines establishing procedural restrictions and laboratory containment conditions. "Biological containment is achieved by using vectors or hosts so crippled by mutation that the organisms bearing recombinant DNA *should be* incapable of surviving under natural conditions

* "Early Days of Genetics," *Heredity,* 4:1–10 (Apr.), 1950.

outside the laboratory." The quote is from *Chemical and Engineering News,* May 30, 1977; the italics are my own. Doubtless, we hope that human organisms bearing recombinant DNA such as the two female subjects of Cline's work do indeed survive, and that they do so without harm and perhaps with benefit from the work performed. The quote is not referring to either this use of genetic engineering or this kind of host. Instead, it refers to the hosts of disease agents, agents propagated in, for instance, *E. coli.*

"In science we always know much less than we believe we do."—Erwin Chargaff*

Escherichia coli, E. coli for short, is a bacterium common to human intestines. In the form of its laboratory derivative it has been as controversially exploited as it has been technically manipulated. That the microbe has been both a potential demon and an acclaimed hero is testimony to the fluctuation in attitudes toward the very technology it innocently promotes. The favorite host for experiments in genetic engineering because it is well understood, readily manipulated, welcoming as a host, and highly available, *E. coli* is the standard laboratory organism. But the human intestine is generally the bacterium's own favorite; both laboratory technicians and the public, then, may be considered vulnerable to *E. coli.* "In the worse conceivable case, an *E. coli* bacterium infected by a virus which some manipulation had unintentionally rendered pathogenic might escape from the laboratory, infect the population outside, and set off a human epidemic of perhaps myxomatosis-like proportions. The incidence of laboratory acquired infections—5000 in the last 30 years, a third of them in laboratories with special containment facilities—suggests that the eventual escape of such an organism, if created, could probably be expected."[5]

E. coli host-vector systems soon began being modified with the goal of minimizing the possibility of unwanted infections of viral and other genes, yet their reliability had not been established as late as 1980, although most genetic engineering projects have continued to rely

* "Uncertainties great, is the gain worth the risk?" *Chemical and Engineering News,* May 30, 1977.

upon them. The concern of the scientific community over the potential risk of transmitting foreign genes from *E. coli* K12 cloning vector systems to other microorganisms, let alone to humans, sparked research into safeguarding measures. Use of *E. coli* as a vector has continued even as the debate of its safety for this use has continued. Yeast, while it has certainly not replaced the bacterium has been, however, becoming more and more popular as a somewhat safer and more efficient host vector.

"There are no nonpolluting technologies. With recombinant DNA, there will be a pollution of a new kind, biological pollution, pollution that grows as the organisms reproduce themselves. Yes, you could say that it's a renewable resource, but the pollution from it is renewable, too."—Jonathan King, MIT

Since the 1920s crude methods for modifying the hereditary message have existed in the form of X rays, mutagens, and irradiation. The methodology for hereditary modification has become a comparatively fine technology of increasing sophistication. Among the methods are those specific to the transfer and immobilization of electrophoretically separated DNA fragments. The fragments are transferred to aryldiazonium derivatized paper (DEM paper). In volume 207 of *Science* W. Scott Chilton reported that the technique could expose workers to a carcinogenic by-product. The concern especially involved two pieces of common laboratory equipment: a dry ice-acetone pump and a fume hood. "Even at 15 mm of pressure," wrote Chilton, "the pumptrap was found to contain a large amount of bischloromethyl ether. Any uncondensed bischloromethyl ether would be pumped into the room. Consequently, the pump should be placed in a fume hood and provided with several convenient condensing traps for the protection of the workers." Hazards are not generally taken lightly by those working closely with them. They remain, however, a very real threat.

In 1978 foot-and-mouth virus escaped a P4 containment facility on Plum Island, New York. The high-security facility has been likened to the biological warfare laboratories that had existed at Fort Detrick, Maryland. It was the only laboratory in the United States permitted to handle the virus. Scientists there were hoping to clone the virus for further study of how it works (and have since been successful at this; vaccination of cloven-hoofed animals against FMD is foreseeable). The virus was kept from spreading during the outbreak largely because the outbreak occurred on an island. The island was restored to normal operations after several anxious weeks.

Proponents of the recombinant DNA controversy experienced further evidence in their favor when Roger Guillemin and his team at Salk Institute made the discovery *New Scientist* described as follows: "An eminent brain scientist and his colleagues have shown that some of the most accepted tests in molecular biology—the immunological methods for establishing the identity of molecules—may not be as exquisitely specific as most people have assumed. In effect, if a scientist relies on these tests alone to identify a substance, his sample may contain molecules other than the substance he thinks he is studying."[6] Table 4–2 describes some of the results arrived at by Guillemin and his team. While the word "assumed" is dangerous in journalism addressed to a semantics-conscious scientific community, the fact that a method is not what it was thought to be remains. Because of the method, the results of specific biochemical researches were threatened by the mistaken identity of proteins. Laboratory reagents are supplied for standard identification of human proteins existing in laboratory systems. Their use may require that antibodies be labeled with radioactivity and counted, or made fluorescent and photographed; or other procedures such as immunoadsorption may be used. Now, however, immunoassays are believed inadequate without the additional knowledge of the composition and structure of the protein subject. Tagging systems, such as immunoadsorption (antibodies embedded in gel column cling to any protein recognized), which have been used for years, are haunted with the possibility that experiments involving them were contaminated with the "wrong" molecules.

Equally if not more alarming have been the "immortal" cells of a cancer patient which for nearly half a century managed to permeate hundreds of international laboratories in which scientists believed they were studying something quite other than the cervical cancer cells of a black woman of Maryland. Lesser threats to research and its accuracy include upsets such as the 1970s mouse calamity. The description of this lesser calamity is surprisingly not unlike that of a minor nuclear plant breakdown: "a valve stuck someplace" and let steam into an air-conditioning coil, and the thermometer "went off the scale." The result was death to approximately 3000 mice involved in multimillion-

Table 4-2. Confused Protein Identities.[a]

β-Endorphin Fragment	IgG Fragment
valine	lysine
serine	serine
leucine	glutamine
threonine	threonine
cysteine	proline
leucine	leucine
valine	valine
lysine	threonine
glycine	leucine
phenylalanine	phenylalanine
tryosine	lysine
proline	asparagine
serine	isoieucine
isoleucine	isoleucine
alanine	lysine
valine	asparagine

[a] Source: *New Scientist,* April 24, 1980.

Table describes a part of beta-endorphin molecule that antibodies were designed to recognize (left). IgG fragment also recognized in Salk team's experiment (right). This small fragment shows six amino acids occupying the same positions in the two proteins.

dollar research. This cannot be attributed to researcher irresponsibility. Typically, bioscientists do not work with steam valves and air-conditioning coils. This particular incident, then, is an example of extenuated responsibility. No one type of individual, whether it be researcher or electrical technician, can be solely linked to the word *hazard,* as hardly a single experiment can be so linked. Collective participation is necessary to every life event. Hazard may be triggered from any number of sources. It is a curious and troublesome fact that an "airtight" atmosphere chokes the best of us and consequently the best of our endeavors, while an "air loose" attitude or environment has been known to produce awesome catastrophe.

If research with the potentially balanced caduceus within the body, the double helix of DNA, is able to provide any harmonizing influence in addition to that evidenced by centuries of healers for whom it has been an insignia, a blessing would surely be realized. Would we be ready?

With any project that is to continue, it is natural to conclude that risk, its presence or potential presence will not be entirely absent; this is clear even when risk is minimized or rechanneled as a consequence of removal from one spectrum to another perhaps less visible one. In terms of its techniques and products, genetic engineering, considered unnatural by not a few professionals and lay persons, remains bound by this apparently natural law. Laboratory outbreaks like the one of FMD virus at Plum Island, undesirable contamination of research cells, and other hazards to innocent and research organisms are not unfamiliar in genetic engineering. An astounding and embarrassing controversy surrounds what are referred to as HeLa cells.

The nickname, HeLa, derives from Henrietta Lacks. In 1951 tissue samples were removed from Mrs. Lacks, who was found to have cervical cancer. The cells have since traveled worldwide in what appears to be their own determination to grow wherever they can. This determination confused numerous international researches in which the technician started out with one kind of cell only to discover that the cell is not that of Caucasian breast cancer, for instance, nor is it any other expected cell, but is actually the cervical cancer cell of a black woman from Baltimore. The petri dish has been overwhelmed by HeLa cells. HeLa cells travel within tiny airborne droplets created by common lab activities such as pipetting, pulling stoppers, and streaking cultures. There are few clues to alert the researcher to HeLa cell contamination. The sudden explosive growth of cell cultures can signify the success of intended coaxing of cell cultures as much as it might signify contamination. Fortunately, the HeLa cells contain four oddly shaped chromosomes unique to them. These markers testify to the validity of any suspicians about a given cell culture. Backup tests include determination of the presence of a rare digestive enzyme peculiar to the black population and carried in Henrietta Lack's biochemistry. Extensive detective research into the HeLa problem was performed by Walter Nelson-Rees, University of California at Berkeley. In spite of the Nelson-Rees work and the attention that HeLa cells received, the fugitive cells had so vigorously traversed the global spectrum of laboratories that they appeared always to be one step ahead of recognition and capture. A new dimension to the validity of progressive research conclusions was the inevitable consequence of the embarrassing environment defined for scientific professionalism by HeLa cell propagation and contamination. A view of polar opposites is brought to mind by HeLa: both poles are nomadic; they are the unwelcome or harmful HeLa cells within many laboratory cultures, and apparently harmless movable elements of noncoding DNA natural to healthy cells. The problem of HeLa cells is concerned more with the quality of research in a still new field than it is with researcher responsibility, which presents yet another problem.

It is horribly easy to justifiably assert that if self-regulation were a human quality of popular employment, no less popular success, we would all have achieved enlightenment without, presumably, ever having to consider the controversy of genetic engineering. The actions of people like Martin Cline can no longer, it may be hoped, blight the quality of lawbreaking per se which has repeatedly portrayed the ability to benefit society enormously as part of the process of change. Some would say that breaking the laws of nature is just such an example of necessary or preferred lawbreaking. To others it is a heinous crime. But these perspectives can be circumstantially dependent.

In spite of imposed regulations and apart from the possibility of a rebel such as Cline or even such as the clone-crazy character described by David Rorvick in his controversial novel, *In His Image: The Cloning of a Man,* accidents do happen and dangers do demand space.

Medical science as an entirely pure or beneficial "good" received its first major blow with the close of World War II and the realization of what Nazi doctors had been performing on inmates of concentration camps. The medical profession sought to prevent the recurrence of this horror by establishing an ethical code for research involving human subjects. In 1949 the first such code appeared. It was the Nuremburg Code which became known in an updated version (1964) as the widely cited Declaration of Helsinki. In more recent years medical ethics has been made available as a course presented to medical students and has even become a required course for second-year medical students at Columbia University.

The ethics course presented to second-year medical students at Columbia emerged out of small "task force" groups comprised mainly of faculty but also including a few graduate students. It soon became apparent that the abstract approach of humanists and the technical language of clinicians needed a clarifying bridge. This bridge was sought in lay involvement and from among interested students of somewhat removed subjects such as anthropology. The sessions evolved into a true course of study developed by a Columbia dean, Dr. Bernard Schoenberg, internist and psychiatrist, who most regrettably died shortly before the program officially started in 1979. The program is intended as an intermediary between the abstract or philosophical and the practical or clinical. Issues are raised but, curiously, no answers are suggested for these are ultimately up to individual physicians and the law. Perhaps in the future possible answers will be explored as a complement to the intellectual exercise of the course. Students hear lectures, view films or videotapes, and form study groups at the end of each classroom session. They view videotapes such as that depicting a small girl afflicted with spina bifida. The child was not expected to survive even the first few years of normal life expectancy but she did survive them, and continued to survive. At the end of the videotape the camera zooms in on the child, who has been lifted into a toy car. She cannot move from the waist down but now she wants to get out of the car and states this repeatedly to the camera at the conclusion of the tape. Immediately the viewer is struck by the possibility that this child may wish she had not in fact survived, a seemingly unlikely if thought-provoking suggestion. Courses in medical ethics not only bridge the language barrier between humanist and clinician, they bridge the world of the medical student with the world outside an educational institution. The students are faced with questions that have been haunting general and professional vehicles of news for at least a decade. They are reminded to think of their futures as those of a profession rather than a business. They may even consider rumours of the remarkable, such as a child with Down's syndrome and leukemia who not only did not die as predicted, but at the age of 12 no longer showed signs of mongolism, having, by unpredicted and not yet understood natural means, acquired a healthy cell system.

In *Ethical Decisions in Medicine,* Howard Brody, Michigan State College of Human Medicine, describes a multitude of controversial case studies such as the following. Live cancer cells were, in the earlier but not remote part of this century, injected into human subjects. The procedure was part of a study of the relationship of immunity to cancer. The subjects were told only that they would be the recipients of "some cells." The true nature of the experiment was withheld from the subjects, who might have responded with "unnecessary anxiety," even developed phobias. The fact that the "routine" procedures to which they were exposed had previously produced what could not be ruled out as metastatic cancer was not mentioned. Clearly, a renegade was at work.

Experimentation has been condoned in both research and practical uses. It may save the life in one case which in another it might destroy. "The doctor who juggles the doses of his patient's antihypertensive medications to see if he can get better control of blood pressure without excessive side effects is experimenting, in that he doesn't know beforehand what dose will be best; but he is experimenting within the context of medical practice, as the patient's benefit is the main goal, and any new knowledge gained would be purely coincidental (i.e., knowledge of wider applicability beyond that one patient's case)."[7] If the physician were not permitted a measure of leeway in decision-making the individualistic characteristics of patients could not be accounted for in the appropriatenes of a particular method for treatment. Determining the extent of that measure of leeway is another problem.

Medicine is a manipulative art based on evidence which is, except in special cases, external by comparison to that important to genetic manipulation. Genetic engineering has been called unnatural, uncertain, as expressing bad values, and as employing unethical means endangering resource quality and allocation. While Joseph Fletcher, Visiting Professor of Medical Ethics at the University of Virginia School of Medicine, has stated that "a test-tube baby would be more human than one resulting from sexual roulette," genetic manipulation has been widely cited as an attempt to replace natural procreation with an artificial manufacturing process inside the laboratory and outside the traditional context of marriage and family. The uncertainty stems from a lay belief that the outcome of genetic engineering is never predictable, that disastrous consequences can never be ruled out. But simply the desire to change the genetic constitution of offspring has been linked to "bad" value judgment and described as either representing or reinforcing unwittingly dangerous social values. Genetic engineering has also been said to employ immoral means to reach otherwise ethically sound ends, the latter yet debatable. Such means include the killing of excess fertilized eggs as a result of their being discarded during the process leading to in vitro fertilization. Finally, concern for our genetic resources urges careful use. The rapid growth of genetic engineering research and hybridization practices brought cries of resource depletion to professional publications of 1980 and 1981.

In Britain the National Collection of Yeast Cultures, used extensively throughout Europe, was threatened with closure unless adequate sponsoring could be found to support it. Individual microbiologists maintain their own small culture collections but have neither the time nor staff to perform the routine testing and distribution of samples that allow, for example, prevention of sending out the wrong organisms. The applicability of yeast to genetic engineering research and the growth of this field complicated the emphasis of culture collections to a degree that some sponsors, such as Britain's Brewing Research Foundation, found intolerable. The interim crises for microbiologists fearing difficult access to culture collections became ones in which biotechnology actually strengthened its foothold by creating an ever more specific envelope in which the necessity of the repositories was emphasized.

The rapid depletion of germ-plasm resources of natural ecosystems recognized in the early 1980s called for the development of another kind of resource program, one reflecting the conservational quality of ecology. "Save the genes" is synonymous with concern for a reduced variability of natural life forms. The concern is not for microorganisms; their genes are adequately maintained because, in part, the organisms are easily and inexpensively grown and stored. Plants and animals require more attention.

Among domestic animals, four breeds of sheep, two of cow, two of draft horse, one of hog, and numerous chicken strains border extinction at this writing. The possibility for obtaining food, fiber energy, and medicines from certain plant and animal products in the future is, therefore, a more complicated issue, one whose tension can be but moderately relieved by the similar benefits proposed by genetic engineering. The AAAS supported a conservation program geared to genes as resources to be saved from extinction. The support arose at a time when speculation about the essentialness of the gene could parallel the implications of a discovery concerning the brain. This discovery involved an exceptionally large head containing the brain of a genius student whose oversized extremity was noticed by a professor strolling the university campus in California. A subsequent report in *Science* (1981) disclosed that the student's brain functioned without the normally necessary prerequisites, being instead an envelope of spinal fluid only minimally equipped at the surface with the usual synapses. The title question of one writer, "Is Your Brain Really Necessary?" might as well be true for the gene, a concept of sheer but reasonable speculation. In genetic engineering and hybridization it may be asked: Do we truly affect, benefit, or endanger the gene or is there something beyond the gene; or even, is the gene the physiological expression of a truer essence? What corollary is there in the gene to express what is in the brain virtual replacement with spinal fluid? The brain filled with spinal fluid indicates (maybe only secondarily so), a larger head size, or so its first evidence submits. Are there in existence double helices or cells, as there are brains, which are larger than usual because they contain material expressing the vastness of the immaterial? Is all our fuss, then, of affecting the brain and its synapses ridiculous? These statements refer to a possibility whose unveiling is of remote foreseeability yet tremendous significance even though neither all cells nor all brains are necessarily built in this fashion. The typical brains and cells naturally continue to be precious resources.

Six main factors affect the adequate management of genetic resources. In a screening effort, the potential value of genetic resources must be estimated. Only 1 percent of higher plants have thus far been screened. Both collection and evaluation of resources are difficult. Those resources not yet collected or evaluated are, quite democratically, valuable until proven otherwise.

The National Plant Genetic Resources Board made a recommendation early in the history of genetic resource concern. In 1978 it stated that major areas of genetic storage be viewed as a "continuum that sets up a gene flow from source to end use." Everything seems to be very organized, perhaps too organized, and this may be considered from more than one viewpoint. For one, to manage genetic resources is a complex and costly affair. Quantity in terms of collected germ plasm is strongly influenced by cost. In fact, cost has been cited as just about the only clear aspect determining how much germ plasm is collected. The question of quantity is greater than concerns over maintaining the viability of germ plasm under proper conditions. A third problem looms over the difficulties in estimating potential resource value or managing genetic resources: How much plant diversity can be lost without disrupting the ecological balances of natural and agricultural systems? Arguments parallel those discussed in Congress for the protection of endangered species. And, again, the extenuating effects of new genetic technologies on genetic variability, on methods for the storage of germ plasm, and on plant vulnerability have not been determined even to the extent of a sufficient working knowledge. Lastly, various economic and social pressures have affected genetic resources in ways that are not always to the benefit of these resources, despite the good intentions of some of these pressures. The Plant Variety Protection Act, for instance, has been severely criticized. It is attributed with having been a primary cause for the planting of uniform varieties and the loss of germ plasm. Individuality, it may seem, is not to be transcended on this crude plane, maintaining the variety inherent to the unique is essential in the material world. Genetic diversity that is lost is quite justifiably considered irreplaceable. When that irreplaceability touches on the crops that become food on dining room tables, as it may as a result of major food crops becoming more vulnerable because of genetic uniformity, the public may speak more loudly than it ever has in the controversy over genetic engineering.

When methods for the reproduction and redesigning of a species are considered, the response is indicative of a highly sensitive and vital issue. The implications of new or proposed methods raise concerns stretching into and beyond morality just as they reach into the lives of individuals and beyond them into species.

"For the first time in all time, a living creature understands its origin and can undertake to design its future."—Robert Sinsheimer, Caltech, as quoted in "Man into Superman, The Promise and Peril of the New Genetics," *TIME* Magazine, April 19, 1971

"If the human species is capable of being in command of its destiny in so many other ways, should it not be capable of engineering its own future in physical and biological terms?"—Robert MacNeil, MacNeil/Lehrer Report, April 7, 1978

"Reshaping life! People who can say that have never understood a thing about life—they have never felt its breath, its heartbeat—however much they have seen or done. They look on it as a lump of raw material that needs to be processed by them, to be ennobled by their touch."—*Dr. Zhivago* by Boris Pasternak

Reshaping life, or recombination as reconstruction of life itself, is not necessarily worthy of rejection. In Buddhism it may be that reconstructive act which, although dangerous, is sacred. However one approaches results, philosophically or practically, it is clear that the odds are in favor of reshaping what has been sown by nature, as every successful plant breeder is well aware. Less clear is exactly what genetic reconstruction implies. Perhaps it is not only the reverse of the spiritual Buddhic reconstruction by virtue of its being material and approached from a materialistic viewpoint, but also is the reverse in terms of its quality and its results, however subtly the nightmare may at first present itself. Possibly, it is the material complement to the sacred act.

Homo futurus, the poetic name of the future human species, is depicted as a molded "creature resembling the Superman of the Nietzschean and Shavian dream—or at least one whose powers will be dramatically different from contemporary man's."[8] The developments toward the reshaping of our biological future are great in quantity and broadly ranging. In the back-to-nature movement, a return is seen. In the fear that evolution of the planet and its life forms may have been derived from a black hole into which they must inevitably return, a search is created. It is a search to do what such an event would achieve (a common element of fear): to recreate the event of our origins. The germ cell may be as far as it is desirable to carry such an explosive proposition. We are, it might seem, as far as we could possibly be from the following views of Paracelsus. The father of modern medicine believed that the "Great Mystery is uncreated" and that "neither does it return or is it brought back to itself . . . but it can return to what it was before the Mystery." Perhaps the cell is one of what Paracelsus called the special mysteries springing out of the Great Mystery. However we may define and redefine the various degrees of our origin, the extent of our ability to recreate the development of our emergence from the cell is portrayed in laboratories conducting research into the recreation of the original "germ" atmosphere which evolved into our world, and in laboratories conducting research into the cell which determines the nature of the evolved biological presence. Out of the latter the expected has already been formed in the shape of recombinations. The prospects for existing life and life forms are far from having been lost in the press toward what are perceived as new and original creations. The demand upon the traditionalist to cope with the new is, however, one of the almost relentless, if instructive, stresses of the twentieth century.

From all the attention paid to encouraging the reproduction reshaping of the human species, a narrow-minded historian of the future, or one with extremely limited access to documentation, might be inclined to think that the twentieth century experienced a threatening gross depopulation rather than an overpopulation of its "governing" species. Actually, this attention seems only natural as a counter action to the subliminally frightful freedom provided by ever new and improved resources for contraception. As one of the many existing, developing, and foreseeable methods for reproduction, cloning has thus far engendered the most controversy.

"Species survive because of genetic heterogeneity. Cloning does away with genetic heterogeneity. We don't want to reproduce in a process that limits genetic heterogeneity, so for a very good biological reason, we shouldn't reproduce by cloning. We should use cloning for experimental procedures to give new insights into cell differentiation, new insight into cancer, new insight into aging, perhaps. But it is an inappropriate way for a frog to reproduce, and it's an inappropriate way for a person to reproduce because . . . we need, we thrive on, we desparately need genetic heterogeneity."—Robert McKinnell, University of Minnesota*

* In an interview on the MacNeil/Lehrer Report, April 7, 1978.

The term "clone" is derived from the Greek word for offshoot, *klon*. The concept itself is of great antiquity. If the practice of cloning is not new to earth's life in general, it would certainly be new to the human species, which does not like to think of itself either as a self-dividing amoeba nor as plant cuttings from which grow identical copies of itself. Is this instinctive self-preservation? We have already seen what the lack of variability can mean in terms of resource depletion among other species.

Because of enormous stumbling blocks in human cloning the scientific community recognizes little controversy for a prospect that remains quite distant. This leaves the burden of concern on the humanists and on the public, who have come to perceive definite value to thinking ahead to a time when the cloning of humans could become possible and even promotable, with a view toward its potential popularity. The various predictions as to when human cloning will reach technical feasibility, such as Watson's speculation of within 50 years of 1970, are often a helpless fuel to nothing more than the fires of alarmist frenzy. For all the conscious and unconscious preparatory efforts of the preceding years, the arrival of human cloning will undoubtedly contain an element of surprise inferior to the fact that, surprise or not, no preparation could be sufficient to meet the appearance of the first human clone with equal force. The issues, however, that comprise this preparation also comprise a respectable portion of the controversy in genetic engineering. The issues in cloning revolve around the central problem of identity.

"There already are such widespread divergences regarding the sacredness of the act of human reproduction that the boring meaninglessness of the lives of many women would be sufficient cause for their willingness to participate in such experimentation [as providing surrogate wombs for clones], be it legal or illegal."—James Watson

The biological investment in sexual reproduction (as compared with asexual replication) is extraordinary. It reveals a measure of its importance to species evolution. "Were human cloning to be accomplished," writes Leon Eisenberg, "it has been assumed that the result would be psychological as well as physical identity between the originator of the clone and his or her cloned offspring."[9] Eisenberg goes on to consider the relationship between predestination and the clone. He writes that the "unwarranted assumption [of complete identity been the clone and its originator] reveals the persistence of the concept of predestination as a causal force in development, despite its incompatibility with biological evidence." Although it cannot be predicted that a clone will indeed acquire the full set of psychological qualities of its donor, the fine lines between nature and nurture do not yet explain away the assumption, if it continues even subliminally, described by Eisenberg.

The early Greek concept about the unfolding of preexisting structures, the doctrine of preformationism, is reflected in the Hippocratic corpus and expressed in poetic form by the Roman, Seneca: "In the seed are enclosed all the parts of the body of the man that shall be formed." Even early microscopists believed that in sperm they could perceive a homunculus. The improvement of microscopic resolution denied the presence of the expected structures and, as Eisenberg writes, "experimental manipulation of embryos produced abnormal monsters which could not have been already present in the seed." Apparently the seed had unforeseen potentials, at least in terms of its manipulation.

In the seventeenth and eighteenth centuries preformationism evolved into predeterminism. Eisenberg describes the latter as being "now restated in the guise of the early (and no longer tenable) versions of genetic theory. Current polemics on the inheritance of intelligence philosophically reiterate the ancient premises of predeterminism and predestination." Aristotle, struggling with the demands of the uniformly timed appearance of structures, developed the view of epigenesis from his research into egg development. His statement on this was the first enunciation not only of epigenesis but also of successive differentiation. Despite the reality of differentiation and its biological implications a concept popular with the public, and thus far incorrect, suggests that not only a germ cell but even a somatic cell could provide the necessary means for cloning.

Cloning has a history of limited success. In Gurdon's well-known experiments with amphibia (1962, 1968) development proceeded to maturity only one time in one hundred. In the more than 10 years since then the cloning of lower animals and organisms has had greater success. The methodological barriers to successfully cloning a human, however, would be formidable. Further, there are at least three objections to human cloning. It would especially have a devastating impact on the diversity of the gene pool if extensively performed. Selection of traits would be inevitable, reflecting knowledge about traits successful in the past; these traits, however, might *not* be adaptive to the future. Furthermore, the phenotype produced might be unduly vulnerable to its environment. As an act of "positive" eugenics, ethical questions are quickly raised by the possibility of human cloning.

"The main public effect of the remarkable clonal frog produced some ten years ago in Oxford by the zoologist John Gurdon has not been awe of the elegant scientific implication of this frog's existence, but fear that a similar experiment might someday be done with human cells. Until recently . . . this foreboding has seemed more like a science fiction scenario than a real problem which the human race has to live with."—James D. Watson (1971)*

Human embryological development is normally secretive. The most crucial steps always occur in the inaccessible female womb where the embryo grows beyond the range of most manipulation. Normal processes do not permit the steps required for the insertion of an adult diploid nucleus from a donor into a human egg whose maternal genetic material has been removed. Given virtually unmanipulative normalcy, "the idea that we might have a world populated by people whose genetic material was identical to that of previously existing people can belong only to the domain of the novelist or moviemaker, not to that of pragmatic scientists who must think only about things which can happen."[10]

The unexpectedly rapid progress of R. G. Edwards and P. S. Steptoe in their research into the conditions for human test-tube conception mean that "human embryological development need no longer be a process shrouded in secrecy [but instead] an event wide-open to a variety of experimental manipulations."[11] Edwards and Steptoe developed many embryos to the eight-cell stage. Some embryos were carried into the blastocyst stage at which successful implantion into the uterus would be possible. Watson immediately asked why a woman would willingly submit to the laparoscopy operation yielding the eggs for test-

* "Moving Toward The Clonal Man, Is this what we want?" *The Atlantic,* May 1971.

tube conception. One reason is that many women are infertile and in vitro growth of the egg to the blastocyst stage is seen as circumventing this infertility (while infertility among males recommends donated sperm). Availability of the technique presents the possibility of hiring unrelated women to carry a given baby to term if the mother is unable to do this herself, an extreme extension of the wet-nurse idea. Women, however, who simply wish to avoid the discomforts of pregnancy might also seek this route; not an encouraging picture for motherhood. The potential of abhorrent misuse of technology is not obviously of stronger weight than the unhappiness of couples unable to have their own children. That in vitro manipulation of human eggs may become general medical practice "within some ten to twenty years" is a necessary assumption, writes Watson (1971). Human cloning would then be of extreme interest and efforts to achieve it extensive "whether legal or illegal." Cell fusion may turn out to be the technique that achieves for humans what the insertion by micropipette of diploid nuclei into enucleated eggs does for frog cloning.

Watson suggested dialogue between scientists and the world's citizens which would educate the latter in the procedure(s) and its implications; he also recommended that legislative bodies form national science policies. These exchanges and policies did occur, evolving throughout the 1970s. Before these developments became apparent Watson asked of his as yet unresponsive audience, "Does this effective silence imply a conspiracy to keep the general public unaware of a potential threat to their basic ways of life?" It was not, perhaps, a serious question, but the two following it did seem to be so. These asked if the silence were motivated by a fear that the general reaction to informative efforts would be "a further damning of all science, thereby decreasing even more the limited money available for pure research"; and if the silence implied that the majority of scientists live a remote existence precluding the capacity to think rationally about anything outside pure science, dismissing practical matters as subjects they need not cope with. A primary reason, however, has been and continues to be that to most biologists the feasibility of the cloning of humans looks remote and risky.

Joshua Lederberg, Stanford University, was one of the first to discuss cloning as a practical matter. He soon saw agitators of the controversy as "crying wolf" at a time when other concerns demanded immediate attention, and diverted his influence instead toward the prevention of wide-scale and irreversible damage to the human genetic material as had begun to occur through increased exposure to man-made mutagenic compounds. New techniques in reproduction remained little explored ethically for some time. For those attentive, the scope of concern expanded.

Steptoe's laparoscopy technique can yield several eggs from a single donor. The eggs are not all necessarily cultured to the blastocyst stage or reimplanted into female bodies. Excess eggs may be disposed of or become likely candidates for a variety of valid experimental purposes. With excess eggs "the temptation to try cloning itself thus will always be close at hand . . . it would be shortsighted to assume that everyone would instinctively recoil from such [a] purpose. Some people may sincerely believe the world desperately needs many copies of really exceptional people if we are to fight our way out of the ever-increasing computer-mediated complexity that makes our individual brains so frequently inadequate."[12] This view has proved itself in William Shockley, as will be seen later.

The problems in reaction to asexually produced children, the societal position of the children are of concern, as are the children's own reaction to their environment, an environment highly pressured by simple desires capable of exceptional means by which to demand their possible realization.

In vitro cell fusion is a necessarily controversial technique, and its applications range well beyond test-tube conception. It is a very good avenue toward understanding the genetic basis of cancer. All over the world, cancer cells are fused with normal cells in an effort to identify the specific chromosomes responsible for different forms of cancer. Fusion techniques also form the basis of many efforts to determine the biochemistry of diseases such as cystic fibrosis and multiple sclerosis. This presents a complicated problem for those persons alarmed by the possibilities of the technique in procedures such as cloning. "Any attempts now to stop such work using the argument that cloning represents a greater threat than a disease like cancer is likely to be considered irresponsible by virtually anyone able to understand the matter."[13]

Before a 1971 meeting of a committee of the House of Representatives Watson presented his dismay over the insufficient alerting of the public to some of the profound implications of genetic research.

His efforts did not end there. Nor did the controversy cease over the new biology to which he had been so instrumental by endeavoring, with Francis Crick, to elucidate the structure of DNA.

As a metaphor dramatizing issues developing in bioethics, cloning is of definitely immediate concern. Biologists, ethicists, and social scientists, writes Willard Gaylin, see it as a "device serving to focus attention on identical problems that arise from less dramatic forms of genetic engineering. . . ."[14] The new technologies are all, for instance, motivated by humanitarianism, except in war research.

Centuries were required to realize that plants have heterosexuality, and centuries more to rediscover and indelibly reaffirm this fact. The rediscovery of plant sexuality has enabled science to foster a broad range of research dependent upon a sexuality rediscovered at a time when sexuality per se was a subject of much controversy for science. For more than the decade to 1980 morality grew into a maturity tested by a recurring staunchness currently expressed in the form of something calling itself the "moral majority." Common clonal, or other laboratory-assisted means of reproduction could mean a safe absence of sexuality for the individual. On the other hand, an oversexed society is envisioned as new genetic, chemical, and electronic sexual enhancers develop. Physician John Taylor has suggested that sex could become so much fun that the popularity of nonsexual activities would be jeopardized. In subjects such as these it is sometimes difficult to appreciate with what objectivity such potentially subjective visions are cast. In context of the frantically alarmed controversy in genetics extremes easily make legitimate claims on the future. If it is, as psychology and experience state, that fear begets the very foe of its attention, society would do well to consider the subject and nature of fears before they rise too high.

For James Watson the subject of cloning is one of the few contentions claimed by the man who signed the letter of moratorial recommendation only to later denounce such conservation and its proponents. Before losing his concern, Watson asks: Do we really want a dozen Einsteins or "multiple likenesses of Raquel Welch?" Maurice Wilkins answers him with a statement whose accuracy the public is unsure of. Wilkins states that "cloning may be possible, all right, but it's not socially likely. Not on any serious scale. One cannot easily see what are the real social pressures that could lead to any wide attempts at cloning humans."[15] If one looks hard enough at the late twentieth century it is indeed possible to discover within it the beginnings of a redefined responsibility and desire to carry through such a responsibility. If this were not so the whole meaning of a remarkable technological dimension of this age, not only in genetics but throughout science, would be lost on society. Because humanity does struggle with the meaning of this dimension the beginnings of an outdatedness and inapplicability may be attributed to the statement, at once narrowly specific, that historical "mankind soon lost interest in Irresistible Grace; but it quickly appreciated the competent engineering which was due to science." (Alfred North Whitehead)[16] The inapplicability of Whitehead's statement for many periods of time does more justice than injustice to science. The intellectual suppleness essential to mythological awareness is the same suppleness that allows the creativity by which great science and also great art are achieved.

Society is witnessing a reawakening to various spiritual philosophies and the creation of new branches of doctrine at the same time in which cloning revitalizes our awareness of physiological identity by proposing the duplication of organisms. Of the many parallels that may undoubtedly be discovered in this coincidence is surely the following: The removal of the nucleus from a cell and transplantation of it to another is like the relationship between certain gurus and their disciples whose indoctrination is served by the transference of whole qualities from the guru's own spiritual "center." It is easy to romanticize the parallel between the cell and the spirtual center into, "As without, so within." Such a parallel actually refers to two polarized, if conceivably mutually supportive, interiors.

"It is very hard to identify a bad gene. It is even harder to identify a good one." Hogland and Burnhoe; Encyclopedia of Bioethics, Vol. II, Reich, Warren T., Free Press, Macmillan, 1978.

A list of 1500 distinguishable human diseases reflects those already known to be genetically determined and is added to by new examples being reported every year. Many genetic diseases among humans are rare. The incidence of phenylketonuria is, for instance, only about one in 18,000 live births, with 200 to 300 cases yearly in the United States. Cystic fibrosis of the pancreas occurs approximately once every 2500 live births. The molecular basis for genetic diseases is not yet understood in most cases. A 1970 review by V. A. McKusick, however, listed 92 human disorders genetically determined by a specific and identified enzyme deficiency.

In a very conscientious article appearing in *Science* Theodore Friedmann and Richard Roblin write that "whether genetic engineering techniques can be developed for therapeutic purposes in human patients without leading to eugenic uses is an important question."[17] This is a subject many authors, including Friedmann and Roblin (who do, however, courageously approach a portion of ethics), feel is "mostly beyond the scope" of their particular endeavor. While the problem is not ignored it is perplexing, and is generally confined to ethicists.

The consequences of a gene mutation which renders an enzyme defective, hence creating a genetic disease, can result in any one of the following natural ways.

- Failure to synthesize required compounds.
- Accumulation of abnormally high concentrations of a particular compound and its further metabolites by other biochemical pathways.
- Failure of properly regulating activities of another enzyme due to loss of the normal feedback inhibitor.
- Failure of a regulatory step in a linked pathway. This may occur because of the absence of certain compounds.

In theory, certain recessively inherited genetic defects which result in a deficiency of a gene product, such as an enzyme or a hormone, might be corrected by gene therapy. The goal of the therapy would be to restore the deficient gene product. Many genetic defects cannot be treated in this way. For instance, neither Marfan's syndrome, showing dominant inheritance, nor those defects caused by chromosome abnormalities, such as Down's syndrome are likely candidates for amelioration by what may be the typical gene therapy.

Dietary therapy, drug therapy, and gene product replacement therapy are usually the routes employed for treating human genetic disease. For instance, diets low in lactose or phenylalanine are used in treating individuals for galactosemia and phenylketonuria, respectively. Drug therapy blocks or reduces accumulated metabolites which are undesired and possibly harmful. Inhibiting the enzyme xanthine oxidase with the drug allopurinol reduces accumulations of uric acid, which is associated with gout and with the Lesch-Nyhan syndrome. Theoretically, the alleviation of certain human genetic diseases might be possible by directly supplying the deficient enzyme, and attempts have been reported. Exogenous enzyme molecules are, however, eventually inactivated or excreted from the body. Enzyme injections would need repetition to manage diseases this way. A similarity to the problem of diabetics is percieved. Many patients would probably begin to form antibodies against the administered enzyme. Insoluble or encapsulated enzyme preparations could provide a means of supplying the enzymes in a more stable and perhaps less immunogenic form (Friedmann and Roblin).

It has become possible to sample cells of growing fetuses in utero. Examination of the cells permits diagnoses of a variety of genetic defects. In their 1972 article Friedmann and Roblin discuss this and write that:

> "If genetic defects are detected, some states will permit an abortion if the prospective parents so desire. We recognize that diagnosis in utero and abortion raise difficult social and ethical problems of their own and cite them only to indicate that there are additional alternatives to prospective gene therapy for coping with human genetic disease."

The conflicts themselves seem to be as numerous as the varieties of genetic conditions with which they are concerned. Many genetic diseases exist which do not respond to treatments. Most genetic disorders of amino acid metabolism, except phenylketonuria, are not, for instance, satisfactorily controlled by dietary therapy. Researchers are not convinced that storage diseases, those associated with lysomal enzyme deficiencies, will be responsive to enzyme therapy, and these diseases are probably not controllable by dietary restriction. Even cases in which disease management is effective seldom present perfection; this is evident in diabetes, where there is an increased incidence of vascular and other disorders. Honing in on the genetic level is stimulating to attempts to get beyond the limitations of current therapies. At this level researchers approach the very site of primary defect.

Modification of specific cellular characteristics by employment of exogenous DNA has been explored. During 1969 and 1970 several supportive discoveries, assertions, and trials were made. These involved the DNA-mediated genetic modification of several different kinds of bacteria. Also included was the experiment-based suggestion that genetic properties of mutant *Drosophila* strains are modifiable by treatment of their eggs with DNA extracted from other strains. There was also the subject of the treatment of human cells in vitro with DNA extracted from SV40, a maneuver producing permanent inheritable alteration of several cellular properties. The proposition that evolved from these and other developments was therapy with the gene itself.

The concept of gene therapy is that permanent and inheritable genetic modification can be created by means of DNA. The theory requires uptake of exogenous DNA from the extracellular environment. It has tremendous potential for a future as a therapy of high repute and as an issue highly controversial in its nature.

Physicians, we are reminded by Friedmann and Roblin, have not always waited for a complete evaluation of new and potentially dangerous therapeutic procedures before putting them to use. Nor are physicians alone in this, as is evidenced by the attempts of Cline as well as of others. History is replete with such attempts. The basic aspects of virology, for instance, were little known during Jenner's development of a vaccine against smallpox.

"There is currently (1971), and there may continue to be, a tendency to use incompletely understood genetic manipulative techniques, borrowed from molecular biology, in clinical settings."[18] The extent to which this occurs in the face of regulations developed to prevent premature application of techniques and the other conditions of their development is difficult to assess.

Gene therapy may either soften or harden the conflicts between individual good and species good. If it could heal diabetics by correcting the genetic causes of diabetes, for instance, the result would be seen as one that benefits both individuals and the human race. "If," as Roger L. S. Linn has pointed out, "genetic therapy should eliminate some present genetic ailments at the cost of severely narrowing the human gene pool, it may make the human race less capable of meeting future situations in which the physical and social environment may be greatly changed. To ascribe an illusory permanence to existing conditions is a common human error. So there is reason to take seriously warnings against any rash introduction of genetic changes that may be irrevocable."[19] This is not to say that the conditions of diabetics are themselves of particular preference on objective grounds; knowledge about which ailments may indeed be objectively beneficial is necessarily difficult to acquire. Linn continues, "only a cruel ethic would preserve genetic illness and suffering on the chance that it might someday prove valuable in a now unforeseeable situation."

The topic of new life forms, including chimeras, is, surprisingly, less taxing than that of gene therapy. This remains true even given the publicity its controversial nature has encouraged. The idea of creation is somehow more accessible than that of healing the already created.

"When equality is equated with interchangeability, individuality is negated and the human person transmuted into a faceless mask."—Ruth Nanda Anshen*

Humankind has spent years looking at intelligence among animals which, if not equated with its own intelligence, may be found at least to be similar. That this similarity of intelligence might allow for an interchangeability between the activities of different species is less immediate than physical interchanges. Nature has spent thousands of years in an effort to ensure a separation between species. Humankind, in fits of debasement and acclaimed endeavors of technological prowess, has endeavored to break the barriers without belief that they could indeed be broken. In the natural environment man has largely failed at this. In the laboratory he has succeeded, with bacteria, plants and higher life forms, in creating hybrids, crosses, chimeras. The above statement from Anshen is true for any form embodying the least degree of individuality. When equality is equated with interchangeability the mask of affinities may be a beautifully apt symbol for that which is shared by individuals. This alludes to that infinite essence which is expressed in the particulate.

The achievements of the new biology referred to as new life forms, combinations of existing life forms, are chimeras. With chimeras certain life forms are equal in that their characteristics, while individualized, are interchangeable. If the point were stretched the attribution

* In her introduction to Werner Heisenberg's *Across the Frontiers,* Harper and Row, 1974.

could be applied to any of the achievements of genetic engineering on the basis of its integrating what was with what might have been in order to create what is, the basis of gene therapy. Determining whether the concept for chimeras was historically developed out of embryonic atrocities, clearly and physiologically beastlike people (hirsutes), developed out of naturally occurring mutations, or any other natural phenomena, or simply anthropomorphism, would do little to assist the debate over a reality largely felt to be quite new. One thing is certain: In no discovered record of antiquity is there a reference or even an allusion to a physical manipulation of life capable of successfully creating new life forms which cross species barriers at the level of the gene.

The chimera created by Ananda Chakrabarty has brought the concept of new life forms forward. The Supreme Court's attempt to comprehend the issues involved in deciding whether or not life is patentable has been pitiable. An air of confidence could only superficially surround the Supreme Court's proceedings. Molecular biology and its new life forms were not about to evaporate, and the nature of concern once familiar to the Asilomar Conference did not take advantage of this opportunity to reemerge in its full glory. Debate was left largely to reporters. What is the meaning of the creation of new forms of life? What threat does a chimera pose? These are difficult questions to answer in the best of times; under the pressure of immediacy they become nearly impossible. The Supreme Court would make a decision. Few if any technical or philosophical concerns regarding chimeras and the concept of new forms of life could be resolved. While it is hardly possible to predict the extent of controversy over the subject had the court not ruled, as it did, in favor of patentability, there is little indication that the scientific community would have been up to a very in-depth and extended debate on a subject whose fundamental awesomeness is almost blinding.

The most exciting nonmedical development of molecular biology is surely this creation of new life forms. Some argue that the forms are not really new at all. Paul Margoland, Public Health Research Institute of the City of New York, finds the very idea of patenting these organisms ridiculous. Chakrabarty's oil-eater, prime subject in the decision making of the Supreme Court, might, declares Margoland, very well exist in nature, concealed by virtue of an instability in natural conditions which results in its dying so fast, generationally, that it is not perceived. As unrealistic as this sounds it may not be entirely untrue. Needless to say it more than borders on the grounds necessary for so outrageous a judgment as that by which "you could patent your children!" The court's decision is not based on the sheer uniqueness substantially if not entirely created by a human act of intervention and manipulation. The fine line determining whether or not children are patentable is not significant in terms of hoards of parents who would not bother with such silliness; it is significant to determining what exactly is a new life or a new life form.

Medicine is a long way from foreseeing its applicability to these so-called new forms of life but may change its attitude in the future, should the potential in this area continue on its course and reach higher forms of and superior methods for created or redesigned life. Whatever shape it takes, in potential or in a perception that may, though it is unlikely, come to realize them as largely metaphorical, the very idea of new forms of life, natural or artificially created, is earth-shaking in its reality. Without it, the theory of evolution could not exist. With it, and in the latter-day sense, we both transgress and materialistically transcend any evolution by intercepting it.

In early 1978, the United States Court of Customs and Patent Appeals overruled the Board of Patent Appeals in favor of patenting a microorganism it referred to as a living invention. Within a year, the case found itself before a higher court and an "inventor," Ananda Chakrabarty, soon became as highly publicized as the new biology. On October 29, 1979, the U.S. Supreme Court announced that it would review the issue of whether or not living organisms can be patented. Although neither of the two cases under review involved recombinant DNA they did reflect the application of unprecedented biological techniques. Upjohn Laboratories put forward a claim on a strain of bacteria producing the antibiotic lincomycin. General Electric made its claim on behalf of the oil-eating *Pseudomonas* bacterium developed by Chakrabarty, who took plasmids from various strains of the bacterium, each imbuing to the original strain the natural ability to degrade a particular crude oil component, and combined them to form a single new strain. It is interesting that society, in regarding the subject of bioengineered life forms, particularly those outside of the medical realm in which humanity frequently expresses a desperate willingness to risk all for the possibility of a cure or measure of relief from its ills, suddenly takes issue with a similar concept which promises other than medical benefits. Media have been replete, for instance, with fictions of automobile gas guzzling by renegade *Pseudomonas.* Questioning the predictablity of the nature of new life forms is but one concern brought to attention by attempts to patent the forms. The prominent issue is in considering the patentability of any life form. The phenomena is the ability to create new forms of life.

The situation was characterized as a narrow question of statutory interpretation "requiring us to construe 35 U.S.C. 101, which provides:

> 'Whoever invents or discovers any new and useful process, machine, manufacture, or composition of matter, or any new and useful improvement thereof, may obtain a patent therefore, subject to the conditions and requirements of this title.'"

Was an engineered microorganism a "manufacture"? Was it a "composition of matter"? Was it covered under the Plant Patent Act?

> ". . . the fact that microorganisms . . . are alive, is a distinction without legal significance."—Court of Appeals

The ultimate decision by the Supreme Court was a defeat for the Federal Government. No new law was invoked. The existing patent laws, the government had stated, did not extend to living things, except for the special case of plants.

The "industrial tools" referred to by the Court of Appeals further in its statement had shown themselves to have very much legal significance. They had been the subject of legal scrutiny since 1972 when General Electric first filed with the Patent Office, and 1974 when Upjohn submitted its case. The cases were not safeguarded by the isolated exclusivity of 1930 and 1970 legislation permitting the patenting of plants. But nowhere else did patent law specifically state that life forms are patentable. So, "what is life?" became a degenerative form of the question returning 40 years after the book of that title broadened biology by conversions it inspired among physicists.

> ". . . the line between life and non-life is so fine as to baffle even the experts. . . ."—Amicus Curiae (in a friend of the court brief filed by the University of California at Berkeley)

"Nature makes so gradual a transition from the inanimate to the animate kingdom that the boundary lines which separate them are indistinct and doubtful."—Aristotle

For those most directly involved in the court's review the problem became one of frustrating complexity: "The argument of those now seeking to patent micro-organisms is that, as previously the case with plant biochemistry, modern biology has so blurred the distinction between animate and inanimate matter as to make the concept of life—in the strict legal sense—almost irrelevant. . . ."[20] According to arguments such as that put forward by the University of California, wrote David Dickson in a 1980 report to *Nature,* "There is nothing special about life that makes it inherently unpatentable. 'We do not see any sound reason for making the distinction between the living and the dead,' wrote Judge Rich in giving the Appeals Court's reasons for approving the patent application." This is an electric statement for philosophy, which would maintain a position not unblinded to technological developments.

Supportively affecting the debate over whether life forms such as Chakrabarty's oil-eater are indeed new, "opponents of patenting man-made organisms . . . felt that despite the intervention of man, bacteria are fundamentally derived from nature and thus are in no way inventions in the usual inanimate sense of the term." This portrayel by the *New York Daily News* is so loose that it might be concluded that no invention is really an invention since its materials ultimately derive from nature. Such generalizing detracts from the mainstream of the argument. Opposition, meanwhile, to the idea that there is no sound reason for distinguishing life, about which there is "nothing special" that makes it unpatentable, from nonlife is dramatically developed by the People's Business Commission (friend of the court): "To justify patenting life organisms, those who seek such patents must argue that life has no 'vital' or sacred property. . . ."

Although the Supreme Court was ill-equipped to handle the monumentality of the case in terms of both its professional and its philosophical concerns, and could not even grasp all of the technical scientific language with which it was forced to grapple, a decision was made. In June 1980, "man-made life forms" became patentable.

More than 100 applications for life-form patents were pending.

Now they could be reviewed in a new way. The patenting of life forms was not going to contain a few isolated incidences. It was not going to stop in 1980. It was going to become a popular protective measure with important potential impacts derived from the court's decision and related policy issues. Is one of these impacts the potential to patent the recombined genes themselves?

". . . food plants have historically been in the public domain. You could own a watermelon, but you could not own the genes of watermelon, the whole strain of watermelon. You couldn't prevent another person from planting watermelon. You couldn't prevent them from selling the seeds. These laws [for patenting life forms] move in the direction of saying that a corporation can prevent anybody from growing, from distributing the seeds of a strain that they have patented. And my own feeling is that life forms, like the air, like the water, are much, much too important for the survival of the human species to be allowed . . . to be privately controlled. That much wisdom is not in [our] hands."—Jonathan King, MIT

Another unsatisfactorily answered question concerns patents, for entities which are indeed later discovered to exist in nature. In September 1981, it was announced that a University of Massachusetts research team headed by George J. Armelagos had discovered evidence in the bones of ancient Sudanese that these people were receiving large doses of tetracycline approximately 14 centuries ago. "Can we presume," writes patent lawyer James J. Burke in a letter to the *New York Times,* "that the owners of those tetracycline patents that are still in force . . . will now dedicate them to the public domain?"

The impacts potential as a result of permitting patents on genetically engineered organisms range from immediate effects on the biotechnological industry, on the patent system, and on academic research, to long-range impacts on genetic diversity and on food supply. Morality and propriety loom over these impacts as the two major policy issues. The morality of patenting living organisms and the propriety of permitting private ownership of "inventions" from publicly funded research can be debated almost endlessly. The debate is not one in which biotechnology companies are eager to become involved. Their own debate focuses on other issues considered in the course of choosing in a given situation to take advantage of patent protection or to simply rely on traditional trade secrecy.

The factors deciding technological protection in industry involve what the product is (an organism or the substance it makes), the product's ability to meet legal requirements necessary to patenting, the likelihood of the product's independent discovery by others, whether or not it is a pioneering invention, its projected commercial life and the potential for its improvement by others were it disclosed in a patent, as well as future plans regarding scientific publication, and the economics of patenting.

Many academicians have become concerned about the effects of permissible life-form patenting and about the commercialization of molecular biology in general. It appcars, however, that this too could be debated for some time to come, that is, until the nature of adaptive results (research to business and business to research) achieves some stability over the years to come since the court's ruling at the turn of the decade.

"Science, with technology, is the only way we have to avoid starvation, disease, and premature death. The misapplication of science and technology is due to the fact that the politics are wrong. Now my own view is that the politics are indeed wrong; but politics and science are so closely interrelated that they can hardly be separated."—Maurice Wilkins[21]

When Harvard University began planning the formation of a genetic engineering corporation in the autumn of 1980 it was met with a score of conflicts. The university's plan was met with scorn from numerous other institutions even though it appeared to be a viable way of achieving greater financial independence and a route, commercialism, others had taken, if on a lesser scale, in the past. The proposal called for a major shift in university policy, and Harvard would have established and held a minority interest in the corporation. The potential for serious conflicts with its traditional academic role led Harvard to withdraw its proposal. The relationship between university and business, however, is not dependent on the formation of such corporations. An interrelationship has already started to become well established on less extreme if still controversial levels. As a consequence, *Science*[22] has warned, university research centers are being depleted because they "cannot compete with business in attracting and keeping talented scientists." The young researchers, financially pressed, tend to avoid innovative projects and their simultaneous risks. Val Fitch, Princeton, states that with students taking, often in lieu of graduate school, positions offering salaries higher than those of their professors, the next generation will not be as well trained even as this one.

"Biotechnology epitomises a field in which scientific knowledge and technical expertise are evolving rapidly, where definite customers have yet to develop in any number; where the research of government departments is divided; where the research will be of interest to a wide range of private industry, public services, etc., and where the principle need for growth is in strategic applied research much of which will not yield results of immediate commercial benefit for perhaps a decade."—Robert Walgate, *Nature,* vol. 283, no. 24, January 1980

In the United Kingdom the customer-contractor principle frustratingly emphasized an "unreal division" between fundamental and applied research. In Sweden private firms got off to a very shaky start amid fears that recombinant DNA research is highly dangerous and not to be conducted in densely populated areas. In France research has gone ahead to include collaborations between the university and business which have so far proved successful, as in the case of the University of Technology at Compiègne. With the springing up of new companies engaged in genetic engineering and their eagerness to raise funds, some firms, such as Biogen, were accused of engaging in "science by press conference," commending research results that were as yet incomplete in order to attract investors.

Sharp fears permeated biotechnology in spite of its entrepreneurial gusto. Some of these spoke of commercially oriented researchers increasing the tensions of a profession already faced with its due measure of competitive secrecy, itself controversial in terms of pure scientific ideals. The same fear addressed the attention of researchers to financial rewards rather than to basic research. Politicians, meanwhile, asked why public tax-supported research should suddenly become private property. Clearly, biotechnology in general did not get off to a smooth start. The potential hazards of genetic engineering and biotechnology in particular strained the wriggling giant to its limits. A report invited by the European Commission on the Hazards of Biotechnology stressed many points of concern including the unwieldy scale of industrial fermenters. The cow, it said, is a wonderful fermenter but one that can become readily infected with pathogens damaging, destroying, or infecting its product, and furthermore it produces only 0.0025 in protein compared to an equal protein weight of microorganisms growing in industrial fermenters. Industrial processes naturally attend a larger realm, one that burdens, even threatens, process control. Two more factors helped to make the field an alarming one. Economically there is a disincentive to using severely disabled strains; and the dangers of employing plant pathogens had received but slight attention. These issues, examples of more extensive concerns, led Drs. K. Sargeant and C. G. T. Evans, Microbiological Research Establishment, Porton, U.K., to conclude the following. Hazards rest in the potential of some microorganisms to infect higher species and cause disease; excepting a few that affect plants and insects dangerous pathogens will not likely be used on a large scale industrially; research is greatly needed on long-range airborne transmission of, especially, plant pathogens. Furthermore, in research of a diagnostic orientation and vaccine production, growth of some highly dangerous human and animal pathogens must continue only on a small scale; such organisms should be listed; their handling should be forbidden without license. Attention, the conclusions continued, should be paid to the fact that the chemistry of a proved and licensed fermentation process can alter in response to genotypic and phenotypic changes in the culture, changes difficult to detect. The rapid detection of small numbers of containment microorganisms within large populations requires research that is sorely needed and highly necessary. Research on the mechanisms of and the potential for phenotypic change is also needed. Biotechnology is too diverse, at least in Europe, argued Sargeant and Evans, to be embraced by a single set of regulations. The most serious danger is in the potential for airborne dissemination of pathogenic organisms, especially to countries in which they are not indigenous. Additionally, the unintended and coincidental establishment of new genotypes during continuous cultivation presents a hazardous potential.

Sargeant and Evans made a suggestion. They suggested a regulatory technique already becoming popular for food enzymes and single-cell protein even as the biohazards report was being prepared around the turn of the decade to 1980. Regulations specified to product type, and worked out with manufacturers as well as in conjunction with governments, were the only workable approach to a vast problem.

Business is not the only unrelated interest involved in genetic engineering. The military is also interested. Biological warfare has the potential to take on new meaning.

Differentiation between the two forms of twentieth-century nuclear power is achieved more easily than differentiation between nuclear and atomic power. The only comparable parallel within biology might be one of nuclear and extranuclear heredity. Both couples, those of technological energy and those of microbiology, share notoriety as controversial tools for purposes deemed constructive as well as for the perpetuation of destructive results.

Conflicting statements decorate the subject of biological warfare with unease. The United States unilaterally renounced all offensive biological warfare in 1969. It continued its program, if modestly, of defensive research in this area. The Biological Weapons Convention of 1975 prohibits development of biological agents. It does not bar research. According to the 1978 Department of Defense report on its chemical and biological warfare program (printed in the *Congressional Record,* July 19, 1979) among bacterial toxins under study were botulinum neurotoxins, anthrax toxins, staphylococcal enterotoxins, also enterotoxins produced by cholera and *Shigella* species, diphtheria toxin, and *Pseudomonas* exotoxin A and exoenzyme S. An outbreak of anthrax at Sverdlovsk, the result of an alleged explosion at a biological warfare plant, worried the State Department. Perhaps Soviet adherence to the 1975 convention forbidding development, production, and stockpiling of biological weapons was not as complete an adherence as expected, or at least hoped. Conflicting actions as well as statements appear to be of an international nature. As concerns rDNA techniques specifically, the Department of Defense (DoD) maintained a cautious stance largely because of public opinion. The presence of *Pseudomonas* exotoxin in the DoD's 1978 report, in spite of its being considered of little potency, illustrates the potential overlap between academic use and military interest: The NIH Recombinant DNA Advisory Committee permitted cloning of the exotoxin A of *Pseudomonas* in certain circumstances; permission was later rescinded but only in terms of cloning it in *E. coli* K12, permissible exclusively under at least the minimal containment, P1 EK1 conditions. The DoD endorsed the NIH guidelines, and yet an official of the Department of Defense reportedly stated that the DoD did not currently (April 1980) use recombinant DNA techniques.

In 1977 M. A. Milstein and L. S. Semejko revealed something of the Soviet orientation and discussed the United States military from their vantage point (*Bulletin of the Atomic Scientists,* February 1977). The report began with the subject of the Soviet Union's proposed ban on the development and production of new types of weapons "of mass destruction."

The creation of "genetic weapons" was described by the authors as considered by the Soviet Union to be a very real possibility in the United States. The possibility and danger of "genetic weapons," they stated, was raised in 1973 at a conference on nucleic acids, "whose delegates insisted on bringing it to the U.S. National Academy of Sciences for consideration."

During the 1960s a five-year military program of genetic research on microorganisms obtained some practical results. In one instance the transformation of a microorganism giving rise to plague resulted in a new strain of this pathogen. In 1974 a report by R. Hammershleg was presented, Milstein and Semejko pointed out, to the American Chemical Society meeting in Los Angeles, giving evidence that the Pentagon was exploring the possibilities of selected use of "genetic weapons." Genetic weaponry can be considered one of a variety of bacteriological weapons banned by international law. Naturally, the authors did not describe any exploration or even desire on the part of the Soviets to develop such weaponry. The semantics of the ban, meanwhile, can permit "genetic weapons" to be qualified as partial synthetics which are formed artificially, the authors asserted, "if one takes into account the successful experiments of obtaining nucleic acid by chemical means, which can artificially bestow lethality on different microorganisms. Those same 'genetic weapons,' examined in a broader context, could be considered a new independent type of weapon capable of mass destruction." This statement, assures the authors' editor, Kosta Tsipis of MIT, is a contradiction of facts of not only the Biological Weapons Convention but also of the understanding of it by the scientific community. The editor cites a statement issued after a special session of the 1976 Pugwash conference, Mühlhausen, a statement whose signatures included that of V. A. Engelhardt, dean of Soviet molecular biologists. The statement read:

"It should be pointed out that there is no basic difference between in vitro and in vivo recombination, a technique suitable for biological weapons technology. We particularly welcome, therefore, the statement by the USA delegation to the CCD [Conference of the Committee on Disarmament] in Geneva on 17 August to the effect that the use of recombinant DNA molecules as a technology for developing weapons clearly falls within the scope of the Biological Weapons Convention of 1972 as prohibited, based upon the explicit language of the Convention as well as its negotiating history, and that such an interpretation is of great importance and one on which doubt cannot be permitted to exist. Subsequently, representatives of the United Kingdom and France endorsed this interpretation at the CCD. We are confident that all signatory governments share this view, and we urge that this fact be made explicitly and publicly known as soon as possible in order to allay any unjustified fears on this score."[23]

How extreme is the difference between scientific and military secrecy? What are the extenuating circumstances altering policies in these areas over the years? Supporters of military funding of basic research addressed a January 1980 AAAS session with emphasis on long-term development of new weapons systems. The visions of biological warfare ranged from the extreme, armies of clones, to the realistic, toxicity carriers or biological modifiers, reminding one of chemical warfare. Nowhere else in history, except perhaps in mythology such as that of the Celts', has warfare proposed to be of such inhuman magnitude, the impossible possible, the incredible mundane without the descriptive support of a sense of surprise maintained with difficulty especially after the Vietnam war. Critics argue that military intervention, providing university research support, endangers the comparative freedom of research whose nature as a tool of destruction would condemn the justification of military funding as a financial tool for universities. The extent of military-funded university research is variable. Military research into biological mechanisms and weaponry, an offensive as well as a defensive program, is likely to long remain an inevitable reality.

Wars are played out on populations regardless of whether their inhabitants are healthy or unhealthy and usually regardless of whether or not desirable or undesirable genes are contained within those populations. The former is the concern of medicine; the latter is addressed by eugenics. Eugenics is the only related topic that encouraged controversy before biology fostered genetic engineering.

"He's howling in the pines at the edge of your fingerprints."—Richard Brautigan, "Cellular Coyote"*

Eugenics, in a sense, affirms what is in its present form known as sociobiology to its adherents: Sociobiology describing a materialism that sees biology as the sole Determiner. Eugenics and sociobiology can overlap in the desire to breed individual organisms to display genes society deems most desirable. This overlap, whether eugenics and sociobiology are recognizably directive or only veiled partners in a drive toward perfection, is seen in efforts to intercept the normal breeding of human beings.

Physicist William Shockley, well known for his efforts to establish a hereditary basis for intelligence, is one of at least five Nobelists who have made donations to a sperm bank organized (1980) by Robert K. Graham, a California industrialist. Graham requested the sperm of Nobelists whose progeny would be conceived by the artificial insemination of bright women. Although society no longer openly avows attempts at forced human breeding, both voluntary and involuntary actions are not unfamiliar to the human population of the late twentieth century.

The history of eugenics may be said to date from those specific primitive societies and from animals whose instincts encourage the destruction of newborns who are weak or untimely (in terms of group overpopulation, etc.). In the history of science, eugenics as the encouragement of the desirable and discouragement of the genetically undesirable is recognized to have stemmed from as far back as Sir Francis Galton. He gave the subject its name and in the early twentieth century, when he was in his eighties, devoted the final decade of his life to forming the eugenics movement in England. In consideration for the

---

intellectual climate at the time and in consideration for the man, this was a truly loving, if naive, act.

"An enthusiasm to improve the race is so noble in its aim that it might well give rise to the sense of a religious obligation."—Galton

Whether or not the means toward improvement are always noble is quite another question. Often the means result in the opposite of their intent. The great anthropologist, Claude Levi-Strauss writes in his *Tristes Tropiques* that:

"Men can coexist on condition that they recognize each other as being all equally, though differently, human, but they can also coexist by denying each other a comparable degree of humanity, and thus establishing a system of subordination."

Biological determinism states that human societies are the specific result of biological forces of which the biological "nature" of the human species is part. This does not allow a great deal of room for the importance of nonbiologically based thought in the formation of societies. Sociobiology, a form of biological determinism, has received much criticism. It was formulated into a massive volume, *Sociobiology: The New Synthesis,* by Edward O. Wilson, and published in 1975. Without criticism of its views the book becomes readily praiseworthy, as has been pointed out by several authors including Robert Morrison, who wrote that "as a carefully selected and synthesized compendium of what we know about the social behavior of animals, it stands alone—almost certainly a landmark, quite possibly a monumental one." Even if operative in conjunction with detachment, it is asserted that, in biological determinism, the pattern of society has arisen out of human biology, that social arrangements are essentially unchangeable though they may be altered; alteration would demand continuous conscious social control because the changed conditions would be "unnatural." Sociobiology suggests that human behavior can be analyzed in terms of separate traits which each have a unique genetic basis and evolutionary justification. It asks for a genetic accounting for the presence of attitudes and abilities; and it appears, as a theory, loose enough to allow for speculations which are generally considered of a useless sort such as that a gene for music was bred into the human race as a result of the cultural importance and repetition of musical practices—this seems almost Lamarkian.

Wilson's book caused an enormous stir addressed to concerns not only for humans but for all life forms. Out of the entire work Wilson dedicated, to the frustration of his readers, only one chapter and scattered paragraphs elsewhere to the subject of humans. "My book, "he later wrote, "was approached tentatively and in a taxonomic rather than a political spirit. The final chapter opens with the passage: 'Let us now consider man in the free spirit of natural history, as though we were zoologists from another planet completing a catalog of social species on Earth. In this macroscopic view the humanities and social sciences shrink to specialized branches of biology; history, biography, and fiction are the research protocols of human ethology; and anthropology and sociology together constitute the sociobiology of a single primate species.'"[24] A lot must have gone on before this end passage to spur the controversy of sociobiology; and indeed it did. The nature and significance of its premises is found with the most immediate pertinence in their reflection, reactions such as the following.

"Wilson joins the long parade of biological determinists whose work has served to buttress the institutions of their society by exonerating them from responsibility for social problems."—Elizabeth Allen et al., letter to *New York Review of Books*

"Can there be any point in renouncing the use of words like mind, or goal-seeking, in the face of phenomena of this kind? . . . I think one is bound to come to the conclusion that the sociobiologists are just 'running scared' of ferocious philosophers."—C. A. Waddington *

Biologists have taken exception to Wilson's work either on purely methodological grounds or on moral or ethical grounds. The greatest sorrow and at once trial blessing is the integrational endeavor Wilson has attempted. Sociobiology, in its present form at least, is simply not a center drawing myriads to its fold but is seen rather more often to force incorporation as though it were the duality of the struggling caduceus

* As quoted by Arthur Caplan in his "Ethics, Evol tion, and the Milk of Human Kindness, "*The Hastings Center Report,* Vol. 6:2, April, 1976.

strangling itself in an overanxious desire to harmonize. If it also engenders this attitude sociobiology will surely do more harm than is deserving. Defensive responses such as the following are not uncommon.

"The greatest and perhaps final challenge to sociobiology will come when it invades the province of the humanities and attempts to incorporate them, as it must or be proved false, into the strictly materialistic theory of evolutionary biology."—D. S. Sade*

A lively view of sociobiology has been provided by Arthur Caplan of Columbia College and of the Institute of Society, Ethics and the Life Sciences. In 1976 Caplan stated that:

"... it is the most blatant of fallacies to attempt to formulate moral prescriptions on the basis of empirical scientific evidence. It seems equally obvious to an equally large number of philosophers and scientists that scientific findings directly influence the sorts of moral and ethical prescriptions held to be valid at any given moment. The notorious feuds, at the beginning of this century, between those who believed ethics to be reducible to the empirical properties discovered by science and those who denied that any such reduction was possible, have produced a schism between these two camps that few participants in either group seem willing to acknowledge, much less bridge. . ."[25]

Biological determinism does not easily explain away behavior. Behavior, whether contemporary or ancestral, is commonly viewed as a result of many factors reflecting various degrees of biological and intellectual states and myriad other factors. Culture is burdened by a determination to qualify the place of behavior in society. Even in an age of extensive world travel, or perhaps partially because of it, we are reminded, as Theodosius Dobzhansky has put it, that differences are not necessarily deficits.

"Equality is often confused with identity, and diversity with inequality. . . . Human equality and inequality are sociological designs, not biological phenomena. Human equality consists of equality before the law, political

* As quoted by Arthur Caplan in his "Ethics, Evolution, and the Milk of Human Kindness," *The Hastings Center Report,* Vol. 6:2, April, 1976.

equality and equality of opportunity. These are human rights that come from religious, ethical, or philosophical premises, not from genes."[26]

UNESCO reflected its recognition of this view in 1952 when it declared the United Nations statement that "Equality of opportunity and equality in law in no way depend, as ethical principles, upon the assertion that human beings are in fact equal in endowment." The distinguished geneticist, Dobzhansky, clarifies this further.

> "Genes determine the intelligence (or stature or weight) of a person only in his particular environment. The trait that actually develops is conditioned by the interplay of the genes with the environment. . . . We must not brand people or professions as elite or common. . . . The presence of rare abilities need not detract from appreciation of more common ones. . . . Moreover, history is moving in this direction."

Before this move of history's is complete, society will have continued, even as it pulls away from the edge, to brink dangerous fallacies. One such dangerous notion has been that an extra Y chromosome in males may develop antisocial or criminal behavior. Fortunately, it is a notion that has passed out of science and into social and medical folklore where it will wreak havoc and finally disappear. There is a relationship between the chromosome and behavior but it is found to be more casual than causal. The genetic basis for behavior is still a new area of research. It has been considered in several areas, not only in antisocial behavior but in the case of schizophrenia. Research into the relationship between heredity, individual genotype, and schizophrenia began in 1916 when the first family study was reported by Rüdin. Evidence does indeed support the reality of such a relationship in which the genotype interacting with environmental factors can give rise to schizophrenia; but qualification of the exact nature of this relationship remains to be done. The subject of a genetic basis for behavior, whether it be pleasant or unpleasant behavior, raises the issue of behavior modification to a potentially new level. Psychology can contain reams of confusions based, even in its efforts to transcend them, on cultural preferences defining acceptable behavior. Public society has at least as many confusions. Goals in any behavior modification techniques employed for other than research purposes are likely to be of a dubious nature. Research progress in this area has been slow.

Genetic screening and counseling developed out of the eugenics movement. They are addressed to preventative medicine for individuals already living, such as the screening for sickle-cell trait, and to those who may be born if the genetic counselor approves the genotypes of the particular potential parents. "The potential advent of widespread genetic screening," states a special 1972 report from a research group of the Hastings Center, "raises new and often unanticipated ethical, psychologic and sociomedical problems for which physicians and the public may be unprepared. To focus attention on the problems of stigmatization, confidentiality, and breaches of individual rights to privacy and freedom of choice in childbearing, we have proposed a set of [guiding] principles."[27] The main principles emphasized by the research group included well-planned program objectives, community involvement in areas immediately affected by screening, provisions for equal access, proper testing procedures, a clearly defined procedure for obtaining informed consent and for maintaining safeguards for the protection of the subjects, and an understanding of the relationship between screening and potential therapy. Within 10 years of the proposal of these principles, which could not have been well heeded, screening for sickle-cell carriers caused ostracism based on misunderstandings that even led the U.S. Military Academy to expel one such carrier simply on the basis of the cadet's genotype. Sickle-cell anemia is a horrendous condition for those actually diseased; but no more so for those surrounding the subject than many other illnesses. Furthermore, not everyone containing the basis for sickle-cell anemia is going to actually become afflicted. But the frenzy of the hunt for the trait created widespread panic and despair among the black population it is peculiar to and among the schools and professions of these individuals. Screening programs obtain information from large numbers of both normal and asymptomatic individuals. Involvement in a screening program, therefore, need not imply more than any of the other numerous query programs of a given population.

"I will follow that system of regimen which according to my ability and judgement, I consider for the benefit of my patients, and abstain from whatever is deleterious and mischievous."—The Hippocratic Oath

It is often difficult to perceive whether a particular system is itself at fault or whether the presentation of that system is the mischievous element in the experience of a program such as genetic screening. Part of genetic control, or attempts at control, is the determination of "fitness" for reproduction and both "fitness" and desirability for birth. Choosing the sex of a child is now an option for expectant parents. Many more options, perhaps even behavior modification for desired characteristics will become available to future parents. On this level genetic control is almost uncontrollable. While the genetic counselor may attempt persuasion, ultimate decision making is based on individual philosophy; unanimous collective philosophy is predominant only in primitive societies or those of great antiquity. The nature of the professional persuasion, however, can be erroneously incriminating as facts are scientifically agreed or disagreed upon.

"Facts cannot be divorced from cultural contexts."—Stephen Jay Gould, "What Color Is a Zebra?" *Natural History,* August 1981.

Eugenics may be seen as an attempt to meet pressing human needs. One of these had been perceived as a large genetically deteriorating situation. The vision of a genetic apocalypse, however, needs qualifying and may no longer apply to humans, as the American Eugenics Society suggested in a six-year report ending in 1970. Other issues are thornier. One such interest is the state interest in preventing the birth of genetically diseased children on the basis that these children often become an economic burden sapping the resources of the state. Involuntary sterilization of "mental defectives," while it may not remain valid in light of subsequent judicial developments, is another example. Actions and interests such as these deny the very nature of the twentieth century. This is a century of great technological achievement on every level including medicine, a century of tremendous promise and hope even amid fears that the manifestations of its hope and promise may not have been presented and developed in ways sufficient to impede an apocalypse of dimensions far exceeding that of a "genetic twilight." Are we simply to deny that problems have continued to exist in a society of promise and hope? Can prevention of reproduction really erase these problems from view? This is not a matter of medical or behavioral genetics, it is a matter of social behavior, and represents a problematical stage in the evolution of that behavior.

"The procedure of sexual enjoyment has finally become wholly divorceable from the procedure of procreation."—Gordon Rattray Taylor

Fortunately, the use of technological means in "divorced" procedures of procreation is limited to particular instances. The weight of those instances, however, may be borne by the whole. Geneticists, as much as other professionals, are fallible human beings. Broad-ranging interdisciplinary and social participation is essential to "trying to lay out the limits of scientific expertise, responsibility, and freedom" (A. L. Caplan).

The nature of reproduction is changing. Initiating life without the traditional male involvement through, for instance, in vitro fertilization has already occurred and is gaining popularity among couples otherwise unable to give birth to their own children. This necessitates a change in the female role as well, but one of far lesser disinvolvement than that of the male, who does not even have the opportunity to carry a baby to term, however that baby has been conceived. Does partial responsibility in conception and the near months proceeding it imply partial responsibility in child rearing? Not at all; but other implications of no less importance do exist. An enormous implication is described by William Irwin Thompson in his *The Time Falling Bodies Take to Light:*

> "The effort to displace the female seems to be the archetypal foundation for civilization, for mankind has been at it ever since, whether he is challenging Mother Nature in flying away from her in rockets, or in changing her on earth through genetic engineering, man has not given up the attempt to take away the mystery of life from the Great Mother and the conservative feminine religion."

However we resolve the male/female, anima/animus, earth/heaven issue in the course of our pursuit of it, it may be necessary to explore the female pole or principle further. This is already encouraged in the women's liberation movement. A coincidental and possibly deleterious, if temporary, result is the interest of many women in avoiding the rigors of childbirth. Fortunately, it may be easily assumed that the foundations for procreation and the traditional methods relating to it are sufficient to preclude widespread misuse of less personal modern techniques including that allowing for conception outside of the parental body. However, given the growing participation in test-tube conception it is no longer fair to refer to test-tube babies as such; as they grow to maturity, these children doubtless find themselves as legitimate a life form as any other child.

"Two hundred years ago," writes David H. Smith, Indiana University, "Rousseau raised an issue that biomedicine has forced on us all: whether there are things we should not know." Knowledge has experienced abuse throughout history, the abuse of scientific knowledge often having catastrophic as well as subtle effects. Nuclear physics, for instance, was abused when taken beyond its original intent, reducing Hiroshima and Nagasaki to rubble. Stephen Stich suggests that if knowledge were too terrible it would not even be gleaned. Smith does not entirely reject this possibility, but he notes that there is sufficient abuse of existing knowledge to cause concern. Perhaps it is not the knowing, he suggests, but rather the doing that should be the focus of attention. The nature of much knowledge is circumstantial and in a state of existing or potential flux. Other knowledge is more long lasting and appears to have touched on the core of its subject. The knowledge at the root of genetic engineering, or much of it, would appear to fit this second category of lasting validity. The doing that revolves around it has initiated its actions by establishing a state of flux. If that doing were interpretive rather than materialistic it might not have created the existing controversy in genetics. As long as that doing is literal and is materialistic the controversy is likely to continue. The subject is too intimate, at the physiological core of every individual and collective life, for this to be otherwise.

Doing and knowledge often spell humanitarianism. Humanitarianism has become twisted by materialism and very often by its own blindness to the true nature of the needs it strives to meet. The sympathy and manipulation inherent to medicine and to medical genetics are, however, being qualified as new analysis turns many of their traditional premises into unfounded responses. Perhaps we are becoming recivilized.

Bacteria are no longer the culprits of disease; their hosts are responsible. Parenting that results directly in a "mamma's boy" or "daddy's girl" is neurotic, not loving. The handicapped are "stifled by 'kindness'"; the aged deserve greater respect and are capable individuals being incorporated back into a society that has long rejected them; and the dying who might be prolonged in existence if not fully in life are permitted to die when this is found to be more realistic than clinging to that which no longer desires to remain in their embrace. Our understanding of what precisely is sympathetic and where sympathy applies are being scrutinized and clarified. The harmonious interchange of opposites allowed to be themselves, expressed in, for instance, riding the tide of pleasant/unpleasant experiences, may be feasible but it may also include a candor. This candor, well known in the endeavors of scientific research releases the frustration of ineffectual kindness, and extends beyond kindness in order to preserve honesty.

"The inner object of the ritual (to win the Treasure) is not the natural process, but the control of nature through the corresponding creative element of man."[28]

If scientific experiment is by its very nature manipulative must, then, the employment of its results also be acts of manipulation? Evidence so far would indicate a positive answer to this question, and little suggests that a problem may lay not in the manipulation itself so much as its excessiveness and awkwardness at integration with the surrounding environment. Intervention in life processes is necessarily empirical. We cannot presume to have the wisdom essential to intervening on the basis of knowledge that will not change. We can, however, work toward an interaction that allows to gain clearer insight and to act according to the realization that the insight is by and large incomplete to the task at hand.

"Our common concern is the very destiny of the human race. For man has now intervened in the process of evolution, a power not given to the pre-Socratics, nor to Aristotle, nor to the Prophets in the East or the West, nor to Copernicus, nor to Luther, Descartes, or Machiavelli. Judgements of value must henceforth direct technological change, for without such values man is divested of his humanity and of his need to collaborate with the very fabric of the universe in order to bestow meaning, purpose, and dignity upon his existence. No time must be lost since the wavelength of change is now shorter than the life-span of man."—Ruth Nanda Anshen[29]

That this collaboration is needed is without doubt. That it has begun is a recurring, modulated, often tenuous reality. That it will continue, hope speaks, is inevitable.

Section C
Reflections and Definitions

Conclusion

> "The predominant factors are no longer those of our scale; we have come to the edge of a world of which we have no preconceptions, and where all our preconceptions must be recast."*

Our preconceptions about what biological life can mean, through applied genetics, must be recast. The new meaning given to biology is extensive. The doors have continued to open for applied genetics, some, such as the door to agriculture, having taken longer to do so than others, but then if the transition happened any faster applied genetics may have seemed to be performing a monumental takeover; instead it has been perfecting an integration into other modern technology and concepts. Interesting twists occur. Laparoscopy, the operation employed to remove eggs that will be fertilized in vitro, is now of use as an operation toward fertility, yet it had been devised originally toward the opposite purpose, before it was realized that the removed eggs could be fertilized outside the body. Medical genetics is the modern half of results of studies in genetic engineering; the other half, biotechnology, is actually of some antiquity—fermentation for instance is no longer recognized simply as fermentation but as the most ancient biotechnological art. Impressive achievements in medical genetics make many of its early achievements seem of equal antiquity. The vistas of the current age open broad-ranging arms reaching beyond the awesome current facts that, for instance, all the globin genes have been cloned, certain

* Thompson, D'Arcy, *On Growth and Form*.

clusters have been mapped, and the complete nucleotide sequences of particular globin genes are known. Genetic engineering has made significant and consistent strides toward reshaping the way we live and our understanding about the life processes.

"Science and technology . . . have a number of distinguishing characteristics which cause special problems or complications. One is ubiquity: they are everywhere. They are at the forefront of social change. They not only serve as agents of change, but provide the tools for analysing social change. They pose, therefore, special challenges to any society seeking to shape its own future and not just to react to change or to the sometimes undesired effects of change."—Guild K. Nichols*

In the discovery of DNA society finds a light kindled from one technology to another in terms of the possible ways in which genetics may be applied. In the reality of DNA a design analogous to configurations found in sacred geometry roots in the history of civilization the significance of the double helix. As a catalyst of change the genetic engineering which arose out of the discovery of DNA recognizes the stability of wholeness in the cell and the extent of variability of which the expression of this wholeness is capable. The discovery of DNA has shaped a future which society has only begun to realize it is entering.

---

*Nichols, Guild K., *Technology on Trial: Public Participation in Decision-Making Related to Science and Technology,* Paris: Organization for Economic Cooperation and Development, 1979.

References

Chapter 1

1. Gordon, Benjamin Lee, *Medicine Throughout Antiquity.*
2. Thompson, William Irwin, *The Time Falling Bodies Take To Light.*
3. Michell, John, *Ancient Metrology: The Dimensions of Stonehenge and of the Whole World as therein symbolised,* 1981: Pentacle Books.
4. Gordon, Benjamin Lee, *Medicine Throughout Antiquity.*
5. Virchow, Rudolph.
6. Olby, Robert C., *Origins of Mendelism.*
7. Thompson, D'Arcy, *On Growth and Form.*
8. Ibid.
9. Dunn, L. C., *A Short History of Genetics.*
10. Wald, George, "The Origin of Life," *Scientific American,* Aug. 1954.
11. Monod, Jacques, *Chance and Necessity.*
12. Ayers, Charlotte J., *Genetics,* Revised Ed.
13. Ibid.
14. Thompson, D'Arcy, op. cit.
15. Yurnis, Jorge J., Sawyer, Jeffrey R., Durham, Kelly, as quoted by Harold M. Schmeck, Jr., "Surprising Similarity Found in Humans and Apes," *New York Times,* July 8, 1980.
16. Gould, Stephen Jay, "Darwin's deceptive memories," *New Scientist,* Feb. 21, 1980.
17. Paracelsus. Reprinted by special arrangement with Shambhala Publications, Inc., 1920 13th Street, Boulder, Colorado 80302. From *The Hermetic and Alchemical Writings of Paracelsus,* Vol. II Edited by Arthur Edward Waite.
18. Schrödinger, Erwin, *What Is Life?*
19. Pasteur, Louis, *Memoir on the Organic Corpuscles Which Exist in the Atmosphere* (1862).
20. Wake, David B., *Science,* vol. 210, no. 4475, December 12, 1980, p. 1239.
21. Bateson, Gregory, *Steps to an Ecology of Mind.*
22. As reproduced by Dunn, L. C., op. cit.

Chapter 2

1. Zeno speaking in Marguerite Yourcenar's *The Abyss*
2. von-Franz, Marie Louise, *Number and Time: Reflections Leading toward a Unification of Depth Psychology and Physics,* Translated by Andrea Dykes, 1974: Northwestern University Press.
3. Judson, Horace Freedland, *The Eighth Day of Creation.*
4. Purce, Jill, *The Mystic Spiral: Journey of the Soul,* 1980: Thames and Hudson Inc., New York.
5. Doolittle, W. F., and Sapienza, C., "Selfish genes, the phenotype paradigm and genome evolution," *Nature,* vol. 284, no. 5757, April 17, 1980.
6. Doolittle and Sapienza, ibid.
7. Orgel, L. E., and Crick, F. H. C., "Selfish DNA: the ultimate parasite," *Nature,* vol. 284, no. 5757, April 17, 1980.
8. Ibid.
9. Chambon, Pierre, "Split Genes," *Scientific American,* May 1981.
10. Ibid.
11. Watson, James, *Molecular Biology of the Gene.*
12. Watson, James, *The Double Helix.*
13. Watson, James, op. cit.
14. Calvert, Paul, "Electrically conducting polymers," *Nature,* vol. 284, March 20, 1980.

Chapter 3

1. Purce, Jill, *The Mystic Spiral.*
2. Hartman, Philip E., and Suskind, Sigmund R., *Gene Action.*
3. Office of Technology Assessment, *Impacts of Applied Genetics.*
4. Genentech: *Science News,* March 29, 1980; Biogen: *Science News,* January 26, 1980.
5. Bawden, Arthru Talbot, *Man's Physical Universe.*
6. Silverberg, Robert, *The Great Doctors.*
7. Milton, John, *Paradise Lost.*
8. Thompson, D'Arcy, *On Growth and Form.*
9. Ibid.
10. National Academy of Sciences, "Genetic Vulnerability of Major Crops," 1971.
11. Dubos, René, *So Human an Animal.*

Chapter 4

1. Editorial Page; New York Times, October 10, 1977.
2. Davis, Bernard D., *Wall Street Journal,* April 5, 1978.

3. "News and Comment: Genetics: Conference Sets Strict Controls to Replace Moratorium," *Science,* vol. 187, March 14, 1975.
4. Jaroff, Leon, et. al., "Man Into Superman: The Promise and Peril of the New Genetics," *Time,* April 19, 1971.
5. "News and Comment," *Science,* March 14, 1975.
6. Wingerson, Lois, "Mistaken identity of proteins threatens biochemists' results," *New Scientist,* April 24, 1980.
7. Brody, Howard, *Ethical Decisions in Medicine.*
8. Jaroff, Leon, et. al., "Man Into Superman: The Promise and Peril of the New Genetics," *Time,* April 19, 1971.
9. Eisenberg, Leon, "The Outcome as Cause: Predestination and Human Cloning," *The Journal of Medicine and Philosophy.* vol. 1, no. 4, 1976.
10. Watson, James, "Moving Toward Clonal Man," *The Atlantic,* May 1971.
11. Ibid.
12. Ibid.
13. Ibid.
14. Gaylin, Willard, "We Have the Awful Knowledge to Make Exact Copies of Human Beings," *The New York Times Magazine,* March 5, 1972.
15. Wilkins, Maurice, as quoted by Horace Freedland Judson as recorded in his *The Eighth Day of Creation.*
16. Whitehead, Alfred North, *Science and the Modern World;* referring to the "ruthless efficiency" of the seventeenth and eigteenth centuries in clearing the world of "muddled thought."
17. Friedmann, Theodore, and Roblin, Richard, "Gene Therapy for Human Genetic Disease?" *Science,* vol. 175 March 3, 1972.
18. Ibid.
19. Linn, Roger L. S., *Encyclopedia of Bioethics,* Macmillan, 1978, p. 524–525.
20. Dickson, David, "Patenting living organisms—how to beat the bug rustlers," *Nature,* vol. 283, January 10, 1980.
21. Wilkins, Maurice, as quoted in Judson's *The Eighth Day of Creation,* p. 97.
22. "Will Biocommerce Ravage Biomedicine?" "News and Comment" dept., *Science,* vol. 210, no. 4474, December 5, 1980.
23. Milstein, M. A., and Semeiko, L. S., "U. S. military 'R & D' through Soviet eyes," *Bulletin of the Atomic Scientists,* February, 1977; originally appeared in Russian, "The U.S.A. and the Issue of New Types of Weapons of Mass Destruction," *Journal USA,* May, 1976.
24. Wilson, E. O., "Academic Vigilantism and the Political Significance of Sociobiology," *Bioscience,* vol. 26, no. 3, March 1976.
25. Caplan, Arthur, "Ethics, Evolution, and the Milk of Human Kindness," Hastings Center Report, vol. 6, no. 2, April 1976.
26. Dobzhansky, Theodosius, "Differences Are not Deficits," *Psychology Today,* vol. 7, no. 7, December 1, 1973.
27. Research Group on Ethical, Social and Legal Issues in Genetic Counseling and Genetic Engineering of the Institute of Society, Ethics and Life Sciences; Marc

Lappé, program director, "Ethical and Social Issues in Screening for Genetic Disease," *New England Journal of Medicine,* May 25, 1972.

28. Neumann, Erich, *The Origins and History of Consciousness.*
29. Anshen, Ruth Nanda. Specified material from *Across the Frontiers* by Werner Heisenberg. Volume Forty-eight of The World Perspective Series edited by Ruth Nanda Anshen. Translated by Peter Heath. Copyright © 1974 by Harper & Row, Publishers, Inc. Reprinted by permission of the publisher.

Bibliography

Adelberg, E. A., *Papers on Bacterial Genetics,* Second Ed., Boston, Toronto: Little, Brown and Company, 1966.

Asimov, Isaac, *The Genetic Code,* New York: The New American Library, 1962.

Avers, Charlotte J., *Genetics,* Revised Ed., New York: D. Van Nostrand Co., 1980.

Avers, Charlotte J., *Cell Biology,* New York: D. Van Nostrand Co., 1976.

Bandman, Elsie L. and Bandman, Bertram, ed., *Bioethics and Human Rights, A Reader for Health Professional,* Boston: Little, Brown and Company, 1978.

Bateson, Gregory, *Steps to an Ecology of Mind,* New York: Ballantine Books, 1972.

Bawden, Arthur Talbot, *Man's Physical Universe,* Revised Edition, New York: The Macmillan Co., 1943.

de Beer, Sir Gavin, *Charles Darwin, A Scientific Biography,* Natural History Library Ed., 1965, New York: Doubleday and Company, Inc., 1964.

Brody, Howard, *Ethical Decisions in Medicine,* Second Edition, Boston: Little, Brown and Company, 1981.

Chedd, Graham, *The New Biology,* New York: Science and Discovery Books, Basic, 1972.

Clark, Paul Franklin *Pioneer Microbiologists of America,* Madison: University of Wisconsin Press, 1961.

Clendening, Logan, *Source Book in Medical History,* New York: Hoeber, 1942

Collard, Patrick, *The Development of Microbiology,* Cambridge: Cambridge University Press, 1976.

Dawkins, Richard, *The Selfish Gene,* Oxford University Press, 1976.

Dixon, Bernard *Beyond the Magic Bullet: The Real Story of Medicine,* Harper and Row, 1978.

Dobzhansky, Theodosius, *Mankind Evolving: The Evolution of the Human Species,* Silliman Memorial Lectures Series, Yale University Press, 1962.

Dunn, L. C., *A Short History of Genetics,* McGraw-Hill, Inc., 1965.

Edmonds, Paul *Microbiology: An Environmental Perspective,* New York: Macmillan, 1978.

Edwards, A. W. *Foundations of Mathematical Genetics,* Cambridge University Press, 1977.

Encyclopedia of Bioethics, Vol. II, Reich, Warren T., Free Press, Macmillan, 1978.

Ghyka, Matila, *The Geometry of Art and Life,* New York: Dover Publications, 1977.

Gordon, Benjamin Lee, *Medicine Throughout Antiquity,* Philadelphia: F. A. Davis Company, 1949.

Gould's Medical Dictionary, 5th Ed., Philadelphia: The Blakiston Co., 1945.

Grobstein, Clifford, *A Double Image of the Double Helix: The Recombinant DNA Debate,* W. H. Freeman and Company, 1979.

Hall, Manley P., *Man, Grand Symbol of the Mysteries,* Los Angeles: The Philosophical Research Society, Inc., 1972.

Hartman, Philip E., and Suskind, Sigmund R., *Gene Action,* Foundations of Modern Genetics Series, Second Ed., Englewood Cliffs: Prentice-Hall, 1969.

Herskowitz, Irwin H., *The Elements of Genetics,* New York: Macmillan, 1979.

Howard, Ted, and Rifkin, Jeremy, *Who Should Play God?,* New York: Dell Publishing Co., 1977.

Jarvis, D. C., *Folk Medicine,* New York: Henry Holt and Company, 1958.

Judson, Horace Freedland, *The Eighth Day of Creation,* New York: Simon and Schuster, 1979.

Kennedy, Donald (Introduction), *The Living Cell: Readings from Scientific American,* San Francisco: W. H. Freeman and Company, 1965.

Monod, Jacques, *Chance and Necessity,* Vintage Books Edition, New York: Random House, 1972.

Mortimer, Charles E., *Chemistry, A Conceptual Approach,* Fourth Edition, New York: D. Van Nostrand Company, 1979.

Neumann, Erich, *The Origins and History of Consciousness,* Princeton: Princeton University Press, 1954.

Office of Technology Assessment, *Impacts of Applied Genetics,* Washington, D.C., 1981.

Olby, Robert C., *Origins of Mendelism,* New York: Schocken Books, 1966.

Olby, Robert C., *The Path to the Double Helix,* Seattle: University of Washington Press, 1974

Porkert, Manfred, The *Theoretical Foundations of Chinese Medicine*, M.I.T. East Asian Science Series, Vol. III, The MIT Press, 1974.

Portugal, Franklin H. and Cohen, Jacks. *A Century of DNA,* MIT Press, 1977.

Du Praw, Ernest J., *Advances in Cell and Molecular Biology,* Academy Press, Vol. I, 1971, Vol. 2, 1972, Vol. 3, 1974.

Rostand, Jean, *Can Man Be Modified?,* Tr. from the French by Jonathan Griffin, New York: Basic Books 1959.

Schaltes, Hoffman, *Plants of the Gods,* McGraw-Hill Book Company, 1979.

Schrödinger, Erwin, *What is Life?, and Mind and Matter,* Combined Ed., Cambridge University Press, 1967.

Shryock, Richard Harrison, *Medicine in America, Historical Essays,* Baltimore: Johns Hopkins University Press, 1966.

Sigerist, Henry Ernest, *A History of Medicine,* Oxford University Press, Vol. 1, 1951 and Vol. 2, 1961.

Silverberg, Robert, *The Great Doctors,* Scholastic Book Services, 1964.

Singer, Charles Joseph, *A Short History of Medicine,* 2d Ed. by E. Ashworth Underwood, New York: Oxford, 1962.

Steiger, Brad, *Medicine Power,* Garden City: Doubleday and Co., 1974.

Stent, Gunther S. and Calendar, Richard *Molecular Genetics: An Introductory Narrative,* W. H. Freeman and Company, 1978.

Stern, Curt, *Principles of Human Genetics,* Second Edition, W. H. Freeman and Company, 1960.

Taylor, Gordon Rattray, *The Biological Time Bomb,* New York: New American Library, 1969.

Theiler, Carl R., *Men and Molecules,* London: George G. Harrap and Co. LTD, 1960.

Thomas, Lewis, *The Lives of a Cell,* New York: Bantam Books, 1974.

Thompson, D'Arcy, *On Growth and Form,* Abridged Edition, Cambridge University Press, 1961.

Considine, Douglas M., Editor *Van Nostrand Reinhold's Scientific Encyclopedia,* Fifth Edition, New York: Van Nostrand Reinhold, 1976.

Waite, Arthur Edward, Editor, *The Hermetic and Alchemical Writings of Paracelsus,* Vol. II, Boulder: Shambala, 1976.

Watson, James, *The Double Helix,* New York: New American Library, 1969.

Watson, James, *Molecular Biology of the Gene,* Second Edition, New York: W. A. Benjamin, Inc., 1970.

Whitehead, Alfred North, *Science and the Modern World,* New York: New American Library of World Literature, Inc., 1925.

Zubay, G. L., *Papers in Bio-chemical Genetics,* Consulting Eds. James D. Ebest and others, New York: Holt, 1968.

Publications:

American Biology Teacher

Bioscience

Chemical and Engineering News

Ethical Issues in Biology and Medicine

Hastings Center Report

The Journal of Medicine and Philosophy

Nature

New England Journal of Medicine

New Scientist

Philosophy of Science

Quarterly Review of Biology

Science

Science News

APPENDICES

Appendix I

In 1944 Erwin Schrödinger's *What Is Life?* began a history of fascination with, especially, molecular biology.* Certainly the gene was an algebraic unit but Schrödinger described its being also a physical substance of comparatively perfected stability, a substance yet capable of expressing tremendous variety. The work drew physicists to biology. Two sections, found in Chapter 5, Delbrück's Model Discussed and Tested, were of especial interest. These are as follows.

THE APERIODIC SOLID

A small molecule might be called "the germ of a solid." Starting from such a small solid germ, there seem to be two different ways of building up larger and larger associations. One is the comparatively dull way of repeating the same structure in three directions again and again. That is the way followed in a growing crystal. Once the periodicity is established, there is no definite limit to the size of the aggregate. The other way is that of building up a more and more extended aggregate without the dull device of repetition. That is the case of the more and more complicated organic molecule in which every atom, and every group of atoms, plays an individual role, not entirely equivalent to that of many others (as is the case in a periodic structure). We might quite properly call that an aperiodic crystal or solid and express our hypothesis by saying: We believe a gene—or perhaps the whole chromosome fibre—to be an aperiodic solid.

[Schrödinger includes a footnote stating: "That it (the chromosome fiber) is highly flexible is no objection; so is a thin copper wire."]

*Reprinted from *What is Life?, and, Mind and Matter,* Combined Ed., by Erwin Schrodinger Cambridge University Press, 1967. *What is Life?* First published 1944. *Mind and Matter* First published 1958. Reprinted by permission of Cambridge University Press.

THE VARIETY OF CONTENTS COMPRESSED IN THE MINIATURE CODE

It has often been asked how this tiny speck of material, the nucleus of the fertilized egg, could contain an elaborate code-script involving all the future development of the organism. A well-ordered association of atoms, endowed with sufficient resistivity to keep its order permanently, appears to be the only conceivable material structure that offers a variety of possible ("isomeric") arrangements, sufficiently large to embody a complicated system of "determinations" within a small spatial boundary. Indeed, the number of atoms in such a structure need not be very large to produce an almost unlimited number of possible arrangements. For illustration, think of the Morse code. The two different signs of dot and dash in well-ordered groups of not more than four allow of thirty different specifications. Now, if you allowed yourself the use of a third sign, in addition to dot and dash, and used groups of not more than ten, you could form 88,572 different "letters"; with five signs and groups up to 25, the number is 372, 529, 029, 846, 191, 405.

It may be objected that the simile is deficient, because our Morse signs may have different composition (e.g. •—— and ••—) and thus they are a bad analogue for isomerism. To remedy this defect, let us pick, from the third example, only the combinations of exactly 25 symbols and only those containing exactly 5 out of each of the supposed 5 types (5 dots, 5 dashes, etc.). A rough count gives you the number of combinations as 62,330,000,000,000, where the zeros on the right stand for figures which I have not taken the trouble to compute.

Of course, in the actual case, by no means "every" arrangement of the group of atoms will represent a possible molecule; moreover, it is not a question of a code to be adopted arbitrarily, for the code-script must itself be the operative factor bringing about the development. But, on the other hand, the number chosen in the example (25) is still very small, and we have envisaged only the simple arrangements in one line. What we wish to illustrate is simply that with the molecular picture of the gene it is no longer inconceivable that the miniature code should precisely correspond with a highly complicated and specified plan of development and should somehow contain the means to put it into question.

Appendix II

The following is the text of the article presented to *Nature* in 1953 (*Nature,* Vol. 171, April 25, 1953) to describe the Watson-Crick proposal for the structure of DNA.

MOLECULAR STRUCTURE OF NUCLEIC ACIDS
A STRUCTURE FOR DEOXYRIBOSE NUCLEIC ACID

We wish to suggest a structure for the salt of deoxyribose nucleic acid (D.N.A.). This structure has novel features which are of considerable biological interest.

A structure for nucleic acid has already been proposed by Pauling and Corey.[1] They kindly made their manuscript available to us in advance of publication. Their model consists of three intertwined chains, with the phosphates near the fibre axis, and the bases on the outside. In our opinion, this structure is unsatisfactory for two reasons: (1) We believe that the material which gives the X-ray diagrams is the salt, not the free acid. Without the acidic hydrogen atoms it is not clear what forces would hold the structure together, especially as the negatively charged phosphates near the axis will repel each other. (2) Some of the van der Waals distances appear to be too small.

Another three-chain structure has also been suggested by Fraser (in the press). In his model the phosphates are on the outside and the bases on the inside, linked together by hydrogen bonds. This structure as described is rather ill-defined, and for this reason we shall not comment on it.

We wish to put forward a radically different structure for the salt of deoxyribose nucleic acid. This structure has two helical chains each coiled round the same axis. We have made the usual chemical assumptions, namely, that each chain consists of phosphate diester groups joining β-D-deoxyribofuranose residues with 3′,5′ linkages. The two chains (but not their bases) are related by a dyad perpendicular to the fibre axis. Both chains follow right-handed helices, but owing to the dyad the sequences of the atoms in the two chains run in opposite directions. Each chain loosely resembles Furberg's[2] model No. 1; that is, the bases are on the inside of the helix and the phosphates on the outside. The configuration of the sugar and the atoms near it is close to Furberg's "standard configuration," the sugar being roughly perpendicular to the attached base. There is a residue on each chain every 3.4 Å. in the *z*-direction. We have assumed an angle of 36° between adjacent residues in the same chain, so that the structure repeats after 10 residues on each chain, that is, after 34 Å. The distance of a phosphorus atom from the fibre axis is 10 Å. As the phosphates are on the outside, cations have easy access to them.

The structure is an open one, and its water content is rather high. At lower water contents we would expect the bases to tilt so that the structure could become more compact.

The novel feature of the structure is the manner in which the two chains are held together by the purine and pyrimidine bases. The planes of the bases are perpendicular to the fibre axis. They are joined together in pairs, a single base from one chain being hydrogen-bonded to a single base from the other chain, so that the two lie side by side with identical *z*-co-ordinates. One of the pair must be a purine and the other a pyrimidine for bonding to occur. The hydrogen bonds are made as follows: purine position 1 to pyrimidine position 1; purine position 6 to pyrimidine position 6.

If it is assumed that the bases only occur in the structure in the most plausible tautomeric forms (that is, with the keto rather than the enol configurations) it is found that only specific pairs of bases can bond together. These pairs are: adenine (purine) with thymine (pyrimidine), and guanine (purine) with cytosine (pyrimidine).

In other words, if an adenine forms one member of a pair, on either chain, then on these assumptions the other member must be thymine; similarly for guanine and cytosine. The sequence of bases on a single chain does not appear to be restricted in any way. However, if only specific pairs of bases can be formed, it follows that if the sequence of bases on one chain is given, then the sequence on the other chain is automatically determined.

It has been found experimentally[3, 4] that the ratio of the amounts of adenine to thymine, and the ratio of guanine to cytosine, are always very close to unity for deoxyribose nucleic acid.

It is probably impossible to build this structure with a ribose sugar in place of the deoxyribose, as the extra oxygen atom would make too close a van der Waals contact.

The previously published X-ray data[5, 6] on deoxyribose nucleic acid are insufficient for a rigorous test of our structure. So far as we can tell, it is roughly compatible with the experimental data, but it must be regarded as unproved until it has been checked against more exact results. Some of these are given in the following communications. We were not aware of the details of the results presented there when we devised our structure, which rests mainly though not entirely on published experimental data and stereochemical arguments.

It has not escaped our notice that the specific pairing we have postulated immediately suggests a possible copying mechanism for the genetic material.

Full details of the structure, including the conditions assumed in building it, together with a set of co-ordinates for the atoms, will be published elsewhere.

We are much indebted to Dr. Jerry Donohue for constant advice and criticism, especially on inter-atomic distances. We have also been stimulated by a knowledge of the general nature of the unpublished experimental results and ideas of Dr. M. H. F. Wilkins, Dr. R. E. Franklin and their co-workers at King's College, London. One of us (J. D. W.) has been aided by a fellowship from the National Foundation for Infantile Paralysis.

J. D. Watson
F. H. C. Crick
Medical Research Council Unit for the
Study of the Molecular Structure of
Biological Systems.
Cavendish Laboratory, Cambridge
April 2.

References

1. Pauling, L., and Corey, R. B., *Nature,* 171, 346 (1953); Proc. U. S. Nat. Acad. Sci., 39, 84 (1953)
2. Furberg, S., Acta Chem. Scand., 6, 634 (1952).
3. Chargraff, E., for references see Zamenhof, S., Brawerman, G., and Chargraff, E., Biochim. et Biophys. Acta, 9, 402 (1952).
4. Wyatt, G. R., J. Gen. Physiol., 36, 201 (1952).
5. Astbury, W. T., Symp. Soc. Exp. Biol. 1, Nucleic Acid, 66 (Camb. Univ. Press, 1947).
6. Wilkins, M. H. F., and Randall, J. T., Biochim. et Biophys. Acta, 10, 192 (1953).

Appendix III

WATSON/CRICK: GENETICAL IMPLICATIONS OF DNA

GENETICAL IMPLICATIONS OF THE STRUCTURE OF DEOXYRIBONUCLEIC ACID

By J. D. WATSON and F. H. C. CRICK

Medical Research Council Unit for the Study of the Molecular Structure of Biological Systems, Cavendish Laboratory, Cambridge

THE importance of deoxyribonucleic acid (DNA) within living cells is undisputed. It is found in all dividing cells, largely if not entirely in the nucleus, where it is an essential constituent of the chromosomes. Many lines of evidence indicate that it is the carrier of a part of (if not all) the genetic specificity of the chromosomes and thus of the gene itself.

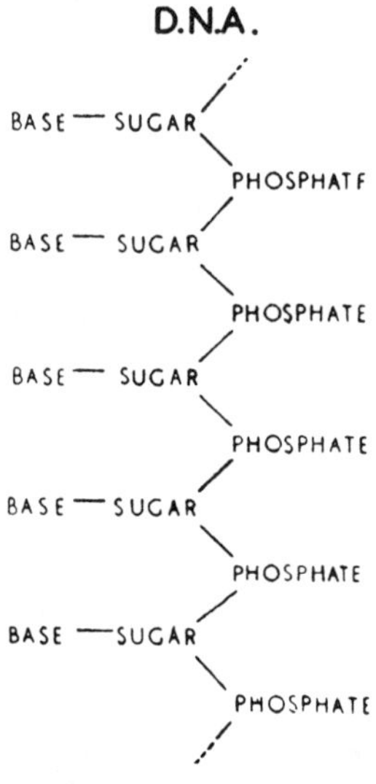

Fig. 1. Chemical formula of a single chain of deoxyribonucleic acid

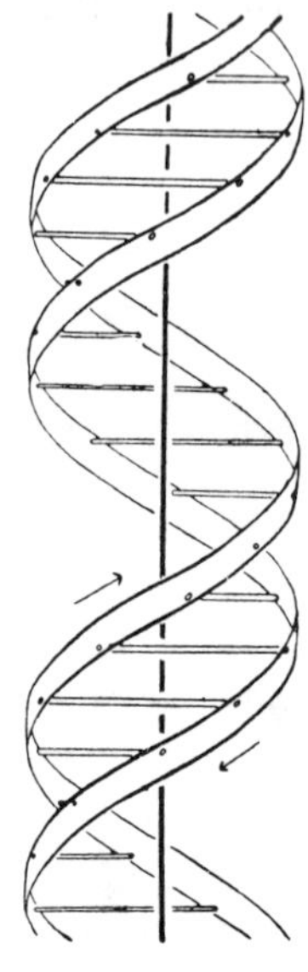

Fig. 2. This figure is purely diagrammatic. The two ribbons symbolize the two phosphate-sugar chains, and the horizontal rods the pairs of bases holding the chains together. The vertical line marks the fibre axis

Reprinted by permission of the authors and the editors of *Nature*. *Nature*, **171**, 964–969 (1953).

Until now, however, no evidence has been presented to show how it might carry out the essential operation required of a genetic material, that of exact self-duplication.

We have recently proposed a structure[1] for the salt of deoxyribonucleic acid which, if correct, immediately suggests a mechanism for its self-duplication. X-ray evidence obtained by the workers at King's College, London[2], and presented at the same time, gives qualitative support to our structure and is incompatible with all previously proposed structures[3]. Though the structure will not be completely proved until a more extensive comparison has been made with the X-ray data, we now feel sufficient confidence in its general correctness to discuss its genetical implications. In doing so we are assuming that fibres of the salt of deoxyribonucleic acid are not artefacts arising in the method of preparation, since it has been shown by Wilkins and his co-workers that similar X-ray patterns are obtained from both the isolated fibres and certain intact biological materials such as sperm head and bacteriophage particles[2,4].

The chemical formula of deoxyribonucleic acid is now well established. The molecule is a very long chain, the backbone of which consists of a regular alternation of sugar and phosphate groups, as shown in Fig. 1. To each sugar is attached a nitrogenous base, which can be of four different types. (We have considered 5-methyl cytosine to be equivalent to cytosine, since either can fit equally well into our structure.) Two of the possible bases—adenine and guanine—are purines, and the other two—thymine and cytosine—are pyrimidines. So far as is known, the sequence of bases along the chain is irregular. The monomer unit, consisting of phosphate, sugar and base, is known as a nucleotide.

The first feature of our structure which is of biological interest is that it consists not of one chain, but of two. These two chains are both coiled around a common fibre axis, as is shown diagrammatically in Fig. 2. It has often been assumed that since there was only one chain in the chemical formula there would only be one in the structural unit. However, the density, taken with the X-ray evidence[2], suggests very strongly that there are two.

The other biologically important feature is the manner in which the two chains are held together. This is done by hydrogen bonds between the bases, as shown schematically in Fig. 3. The bases are joined together in pairs, a single base from one chain being hydrogen-bonded to a single base from the other. The important point is that only certain pairs of bases will fit into the structure. One member of a pair must be a purine and the other a pyrimidine in order to bridge between the two chains. If a pair consisted of two purines, for example, there would not be room for it.

We believe that the bases will be present almost entirely in their most probable tautomeric forms. If this is true, the conditions for forming hydrogen bonds are more restrictive, and the only pairs of bases possible are :

adenine with thymine ;
guanine with cytosine.

The way in which these are joined together is shown in Figs. 4 and 5. A given pair can be either way round. Adenine, for example, can occur on either chain ; but when it does, its partner on the other chain must always be thymine.

This pairing is strongly supported by the recent analytical results[5], which show that for all sources of deoxyribonucleic acid examined the amount of adenine is close to the amount of thymine, and the amount of guanine close to the amount of cytosine, although the cross-ratio (the ratio of adenine to guanine) can vary from one source to another. Indeed, if the sequence of bases on one chain is irregular, it is difficult to explain these analytical results except by the sort of pairing we have suggested.

The phosphate-sugar backbone of our model is completely regular, but any sequence of the pairs of

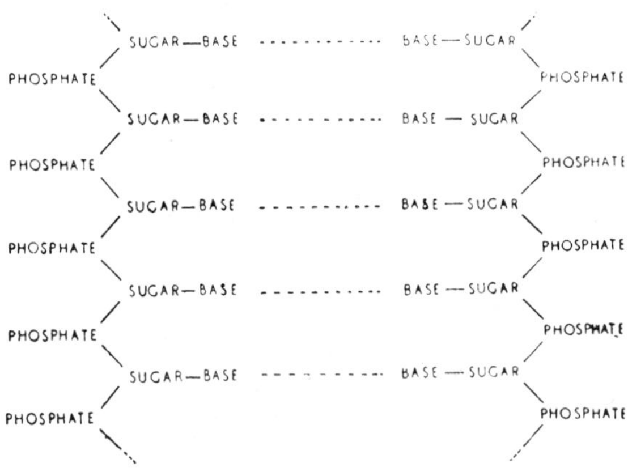

Fig. 3. Chemical formula of a pair of deoxyribonucleic acid chains. The hydrogen bonding is symbolized by dotted lines

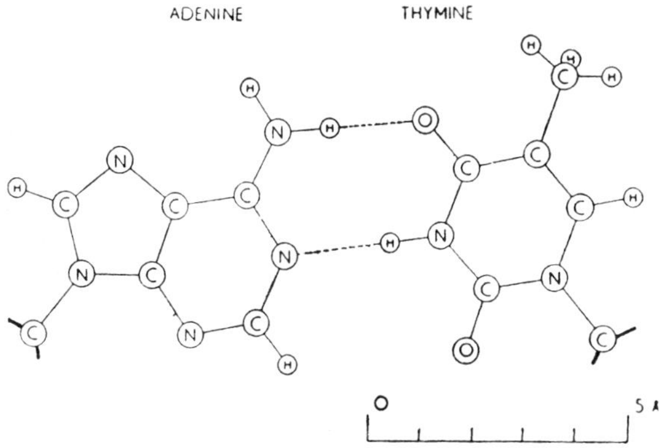

Fig. 4. Pairing of adenine and thymine. Hydrogen bonds are shown dotted. One carbon atom of each sugar is shown

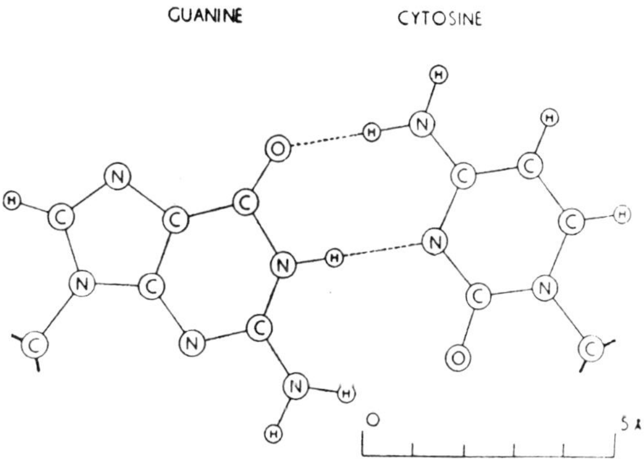

Fig. 5. Pairing of guanine and cytosine. Hydrogen bonds are shown dotted. One carbon atom of each sugar is shown

bases can fit into the structure. It follows that in a long molecule many different permutations are possible, and it therefore seems likely that the precise sequence of the bases is the code which carries the genetical information. If the actual order of the bases on one of the pair of chains were given, one could write down the exact order of the bases on the other one, because of the specific pairing. Thus one chain is, as it were, the complement of the other, and it is this feature which suggests how the deoxyribonucleic acid molecule might duplicate itself.

Previous discussions of self-duplication have usually involved the concept of a template, or mould. Either the template was supposed to copy itself directly or it was to produce a 'negative', which in its turn was to act as a template and produce the original 'positive' once again. In no case has it been explained in detail how it would do this in terms of atoms and molecules.

Now our model for deoxyribonucleic acid is, in effect, a *pair* of templates, each of which is complementary to the other. We imagine that prior to duplication the hydrogen bonds are broken, and the two chains unwind and separate. Each chain then acts as a template for the formation on to itself of a new companion chain, so that eventually we shall have *two* pairs of chains, where we only had one before. Moreover, the sequence of the pairs of bases will have been duplicated exactly.

A study of our model suggests that this duplication could be done most simply if the single chain (or the relevant portion of it) takes up the helical configuration. We imagine that at this stage in the life of the cell, free nucleotides, strictly polynucleotide precursors, are available in quantity. From time to time the base of a free nucleotide will join up by hydrogen bonds to one of the bases on the chain already formed. We now postulate that the polymerization of these monomers to form a new chain is only possible if the resulting chain can form the proposed structure. This is plausible, because steric reasons would not allow nucleotides 'crystallized' on to the first chain to approach one another in such a way that they could be joined together into a new chain, unless they were those nucleotides which were necessary to form our structure. Whether a special enzyme is required to carry out the polymerization, or whether the single helical chain already formed acts effectively as an enzyme, remains to be seen.

Since the two chains in our model are intertwined, it is essential for them to untwist if they are to separate. As they make one complete turn around each other in 34 A., there will be about 150 turns per million molecular weight, so that whatever the precise structure of the chromosome a considerable amount of uncoiling would be necessary. It is well known from microscopic observation that much coiling and uncoiling occurs during mitosis, and though this is on a much larger scale it probably reflects similar processes on a molecular level. Although it is difficult at the moment to see how these processes occur without everything getting tangled, we do not feel that this objection will be insuperable.

Our structure, as described[1], is an open one. There is room between the pair of polynucleotide chains (see Fig. 2) for a polypeptide chain to wind around the same helical axis. It may be significant that the distance between adjacent phosphorus atoms, 7·1 A., is close to the repeat of a fully extended polypeptide chain. We think it probable that in the sperm head, and in artificial nucleoproteins, the polypeptide chain occupies this position. The relative weakness of the second layer-line in the published X-ray pictures[3a,4] is crudely compatible with such an idea. The function of the protein might well be to control the coiling and uncoiling, to assist in holding a single polynucleotide chain in a helical configuration, or some other non-specific function.

Our model suggests possible explanations for a number of other phenomena. For example, spontaneous mutation may be due to a base occasionally occurring in one of its less likely tautomeric forms. Again, the pairing between homologous chromosomes at meiosis may depend on pairing between specific bases. We shall discuss these ideas in detail elsewhere.

For the moment, the general scheme we have proposed for the reproduction of deoxyribonucleic acid must be regarded as speculative. Even if it is correct, it is clear from what we have said that much remains to be discovered before the picture of genetic duplication can be described in detail. What are the polynucleotide precursors ? What makes the pair of chains unwind and separate ? What is the precise role of the protein ? Is the chromosome one long pair of deoxyribonucleic acid chains, or does it consist of patches of the acid joined together by protein ?

Despite these uncertainties we feel that our proposed structure for deoxyribonucleic acid may help to solve one of the fundamental biological problems—the molecular basis of the template needed for genetic replication. The hypothesis we are suggesting is that the template is the pattern of bases formed by one chain of the deoxyribonucleic acid and that the gene contains a complementary pair of such templates.

One of us (J. D. W.) has been aided by a fellowship from the National Foundation for Infantile Paralysis (U.S.A.).

[1] Watson, J. D., and Crick, F. H. C., *Nature*, **171**, 737 (1953).

[2] Wilkins, M. H. F., Stokes, A. R., and Wilson, H. R., *Nature*, **171**, 738 (1953). Franklin, R. E., and Gosling, R. G., *Nature*, **171**, 740 (1953).

[3] (*a*) Astbury, W. T., Symp. No. 1 Soc. Exp. Biol., 66 (1947). (*b*) Furberg, S., *Acta Chem. Scand.*, **6**, 634 (1952). (*c*) Pauling, L., and Corey, R. B., *Nature*, **171**, 346 (1953); *Proc. U.S. Nat. Acad. Sci.*, **39**, 84 (1953). (*d*) Fraser, R. D. B. (in preparation).

[4] Wilkins, M. H. F., and Randall, J. T., *Biochim. et Biophys. Acta*, **10**, 192 (1953).

[5] Chargaff, E., for references see Zamenhof, S., Brawerman, G., and Chargaff, E., *Biochim. et Biophys. Acta*, **9**, 402 (1952). Wyatt, G. R., *J. Gen. Physiol.*, **36**, 201 (1952).

Appendix IV

The "Berg letter," formative in the early beginnings of the controversy from a scientific standpoint, is as follows

POTENTIAL BIOHAZARDS OF RECOMBINANT DNA MOLECULES*

Recent advances in techniques for the isolation and rejoining of segments of DNA now permit construction of biologically active recombinant DNA molecules in vitro. For example, DNA restriction endonucleases, which generate DNA fragments containing cohesive ends especially suitable for rejoining, have been used to create new types of biologically functional bacterial plasmids carrying antibiotic resistance markers (1) and to link *Xenopus laevis* ribosomal DNA to DNA from a bacterial plasmid. This latter recombinant plasmid has been shown to replicate stably in *Escherichia coli* where it synthesizes RNA that is complementary to *X. laevis* ribsomal DNA (2). Similarly, segments of *Drosophila* chromosomal DNA have been incorporated into both plasmid and bacteriophage DNA's to yield hybrid molecules that can infect and replicate in *E. coli* (3).

Several groups of scientists are now planning to use this technology to create recombinant DNA's from a variety of other viral, animal, and bacterial sources. Although such experiments are likely to facilitate the solution of important theoretical and practical biological problems, they would also result in the creation of novel types of infectious DNA elements whose biological properties cannot be completely predicted in advance.

There is serious concern that some of these artificial recombinant DNA molecules could prove biologically hazardous. One potential hazard in current experiments derives from the need to use a bacterium like *E. coli* to clone the recombinant DNA molecules and to amplify their number. Strains of *E. coli* commonly reside in the

*Reprinted, with permission, from *Science,* vol. 185, no. 4148, July 26, 1974, p. 303. Copyright 1974 by the American Association for the Advancement of Science.

human intestinal tract, and they are capable of exchanging genetic information with other types of bacteria, some of which are pathogenic to man. Thus, new DNA elements introduced into *E. coli* might possibly become widely disseminated among human, bacterial, plant, or animal populations with unpredictable effects.

Concern for these emerging capabilities was raised by scientists attending the 1973 Gordon Research Conference on Nucleic Acids (4), who requested that the National Academy of Sciences give consideration to these matters. The undersigned members of a committee, acting on behalf of and with the endorsement of the Assembly of Life Sciences of the National Research Council on this matter, propose the following recommendations.

First, and most important, that until the potential hazards of such recombinant DNA molecules have been better evaluated or until adequate methods are developed for preventing their spread, scientists throughout the world join with the members of this committee in voluntarily deferring the following types of experiments.

Type 1: Construction of new, autonomously replicating bacterial plasmids that might result in the introduction of genetic determinants for antibiotic resistance or bacterial toxin formation into bacterial strains that do not at present carry such determinants; or construction of new bacterial plasmids containing combinations of resistance to clinically useful antibiotics unless plasmids containing such combinations of antibiotic resistance determinants already exist in nature.

Type 2: Linkage of all or segments of the DNA's from oncogenic or other animal viruses to autonomously replicating DNA elements such as bacterial plasmids or other viral DNA's. Such recombinant DNA molecules might be more easily disseminated to bacterial populations in humans and other species, and thus possibly increase the incidence of cancer or other diseases.

Second, plans to link fragments of animal DNA's to bacterial plasmid DNA or bacteriophage DNA should be carefully weighed in light of the fact that many types of animal cell DNA's contain sequences common to RNA tumor viruses. Since joining of any foreign DNA to a DNA replication system creates new recombinant DNA molecules whose biological properties cannot be predicted with certainty, such experiments should not be undertaken lightly.

Third, the director of the National Institutes of Health is requested to give immediate consideration to establishing an advisory committee charged with (i) overseeing an experimental program to evaluate the potential biological and ecological hazards of the above types of recombinant DNA molecules; (ii) developing procedures which will minimize the spread of such molecules within human and other populations; and (iii) devising guidelines to be followed by investigators working with potentially hazardous recombinant DNA molecules.

Fourth, an international meeting of involved scientists from all over the world should be convened early in the coming year to review scientific progress in this area and to further discuss appropriate ways to deal with the potential biohazards of recombinant DNA molecules.

The above recommendations are made with the realization (i) that our concern is based on judgments of potential rather than demonstrated risk since there are few available experimental data on the hazards of such DNA molecules and (ii) that adherence to our major recommendations will entail postponement or possibly abandonment of certain types of scientifically worthwhile experiments. Moreover, we are aware of many theoretical and practical difficulties involved in evaluating the human hazards of such recombinant DNA Molecules. Nonetheless, our concern for the possible unfortunate consequences of indiscriminate application of these techniques motivates us to urge all scientists working in this area to join us in agreeing not to initiate experiments of types 1 and 2 above until attempts have been made to evaluate the hazards and some resolution of the outstanding questions has been achieved.

Paul Berg, Chairman.
David Baltimore
Herbert W. Boyer
Stanley N. Cohen
Ronald W. Davis
David S. Hogness
Daniel Nathans
Richard Roblin
James D. Watson
Sherman Weissman
Norton D. Zinder.
Committee on Recombinant DNA Molecules Assembly of Life Sciences,
National Research Council,
National Academy of Sciences,
Washington, D.C. 20418.

REFERENCES AND NOTES

1. S. N. Cohen, A. C. Y. Chang, H. Boyer, R. B. Helling, *Proc. Natl. Acad. Sci. U.S.A.*, vol. 70, no. 3240 (1973); A. C. Y. Chang and S. N. Cohen, ibid., vol. 71, no. 1030 (1974).
2. J. F. Morrow, S. N. Cohen, A. C. Y. Chang, H. Boyer, H. M. Goodman, R. B. Helling, ibid., in press.
3. D. S. Hogness, unpublished results; R. W. Davis, unpublished results; H. W. Boyer, unpublished results.
4. M. Singer and D. Soll, *Science*, vol. 181, no. 1114 (1973).

Appendix V

ASILOMAR CONFERENCE ON RECOMBINANT DNA MOLECULES*

Paul Berg, David Baltimore, Sydney Brenner, Richard O. Roblin III, Maxine F. Singer.

I. Introduction and General Conclusions

This meeting was organized to review scientific progress in research on recombinant DNA molecules and to discuss appropriate ways to deal with the potential biohazards of this work. Impressive scientific achievements have already been made in this field, and these techniques have a remarkable potential for furthering our understanding of fundamental biochemical processes in pro- and eukaryotic cells. The use of recombinant DNA methodology promises to revolutionize the practice of molecular biology. Although there has as yet been no practical application of the new techniques, there is every reason to believe that they will have significant practical utility in the future.

Of particular concern to the participants at the meeting was the issue of whether the pause in certain aspects of research in this area, called for by the Committee on Recombinant DNA Molecules of the National Academy of Sciences in the letter published in July 1974 (1), should end, and, if so, how the scientific work could be undertaken with minimal risks to workers in laboratories, to the public at large, and to the animal and plant species sharing our ecosystems.

The new techniques, which permit combination of genetic information from very different organisms, place us in an area of biology with many unknowns. Even in the present, more limited conduct of research in this field, the evaluation of potential biohazards has proved to be extremely difficult. It is this ignorance that has compelled us

*Summary statement of the report submitted to the Assembly of Life Sciences of the National Academy of Sciences and approved by its Executive Committee on 20 May 1975.

to conclude that it would be wise to exercise considerable caution in performing this research. Nevertheless, the participants at the Conference agreed that most of the work on construction of recombinant DNA molecules should proceed, provided that appropriate safeguards, principally biological and physical barriers adequate to contain the newly created organisms, are employed. Moreover, the standards of protection should be greater at the beginning and modified as improvements in the methodology occur and assessments of the risks change. Furthermore, it was agreed that there are certain experiments in which the potential risks are of such a serious nature that they ought not to be done with presently available containment facilities. In the longer term serious problems may arise in the large-scale application of this methodology in industry, medicine, and agriculture. But it was also recognized that future research and experience may show that many of the potential biohazards are less serious and/or less probable than we now suspect.

II. Principles Guiding the Recommendations and Conclusions

Although our assessments of the risks involved with each of the various lines of research on recombinant DNA molecules may differ, few, if any, believe that this methodology is free from any risk. Reasonable principles for dealing with these potential risks are: (i) that containment be made an essential consideration in the experimental design and (ii) that the effectiveness of the containment should match, as closely as possible, the estimated risk. Consequently, whatever scale of risks is agreed upon, there should be a commensurate scale of containment. Estimating the risks will be difficult and intuitive at first, but this will improve as we acquire additional knowledge; at each stage we shall have to match the potential risk with an appropriate level of containment. Experiments requiring large-scale operations would seem to be riskier than equivalent experiments done on a small scale and therefore require more stringent containment procedures. The use of cloning vehicles or vectors (plasmids, phages) and bacterial hosts with a restricted capacity to multiply outside of the laboratory would reduce the potential biohazard of a particular experiment. Thus, the ways in which potential biohazards and different levels of containment are matched may vary from time to time, particularly as the containment technology is improved. The means for assessing and balancing risks with appropriate levels of containment will need to be reexamined from time of time. Hopefully, through formal and informal channels of information within and between nations of the world, the way in which potential biohazards and levels of containment are matched would be consistent.

Containment of potentially biohazardous agents can be achieved in several ways. The most significant contribution to limiting the spread of the recombinant DNA's is the use of biological barriers. These barriers are of two types: (i) fastidious bacterial hosts unable to survive in natural environments and (ii) non-transmissible and equally fastidious vectors (plasmids, bacteriophages, or other viruses) able to grow only in specified hosts. Physical containment, exemplified by the use of suitable hoods or, where applicable, limited access or negative pressure laboratories, provides an additional factor of safety. Particularly important is strict adherence to good microbiological practices which, to a large measure, can limit the escape of organisms from the experimental situation and thereby increase the safety of the operation. Consequently, education and training of all personnel involved in the experiments is essential to the effectiveness of all containment measures. In practice, these different means of containment will complement one another and documented substantial improvements in the ability to restrict the growth of bacterial hosts and vectors could permit modifications of the complementary physical containment requirements.

Stringent physical containment and rigorous laboratory procedures can reduce but not eliminate the possibility of spreading potentially hazardous agents. Therefore, investigators relying upon "disarmed" hosts and vectors for additional safety must rigorously test the effectiveness of these agents before accepting their validity as biological carriers.

III. Recommendations for Matching Types of Containment with Types of Experiments

No classification of experiments as to risk and no set of containment procedures can anticipate all situations. Given our present uncertainties about the hazards, the parameters proposed here are broadly conceived and meant to provide provisional guidelines for investigators and agencies concerned with research on recombinant DNA's. However, each investigator bears a responsibility for determining whether, in his particular case, special circumstances warrant a higher level of containment than is suggested here.

A. Types of Containment. 1. Minimal risk. This type of containment is intended for experiments in which the biohazards may be accurately assessed and are expected to be minimal. Such containment can be achieved by following the operating procedures recommended for clinical microbiological laboratories. Essential features of such facilities are no drinking, eating, or smoking in the laboratory, wearing laboratory coats in the work area, the use of cotton-plugged pipettes or preferably mechanical pipetting devices, and prompt disinfection of contaminated materials.

2. Low risk. This level of containment is appropriate for experiments which generate novel biotypes but where the available information indicates that the recombinant DNA cannot alter appreciably the ecological behavior of the recipient species, increase significantly its pathogenicity, or prevent effective treatment of any resulting infections. The key features of this containment (in addition to the minimal procedures mentioned above) are a prohibition of mouth pipetting, access limited to laboratory personnel, and the use of biological safety cabinets for procedures likely to produce aerosols (for example, blending and sonication). Though existing vectors may be used in conjunction with low-risk procedures, safer vectors and hosts should be adopted as they become available.

3. Moderate risk. Such containment facilities are intended for experiments in which there is a probability of generating an agent with a significant potential for pathogenicity or ecological disruption. The principal features of this level of containment, in addition to those of the two preceding classes, are that transfer operations

should be carried out in biological safety cabinets (for example, laminar flow hoods), gloves should be worn during the handling of infectious materials, vacuum lines must be protected by filters, and negative pressure should be maintained in the limited access laboratories. Moreover, experiments posing a moderate risk must be done only with vectors and hosts that have an appreciably impaired capacity to multiply outside of the laboratory.

4. High risk. This level of containment is intended for experiments in which the potential for ecological disruption or pathogenicity of the modified organism could be severe and thereby pose a serious biohazard to laboratory personnel or the public. The main features of this type of facility, which was designed to contain highly infectious microbiological agents, are its isolation from other areas by air locks, a negative pressure environment, a requirement for clothing changes and showers for entering personnel, and laboratories fitted with treatment systems to inactivate or remove biological agents that may be contaminants in exhaust air and liquid and solid wastes. All persons occupying these areas should wear protective laboratory clothing and shower at each exit from the containment facility. The handling of agents should be confined to biological safety cabinets in which the exhaust air is incinerated or passed through Hepa filters. High-risk containment includes, in addition to the physical and procedural features described above, the use of rigorously tested vectors and hosts whose growth can be confined the laboratory.

B. Types of Experiments. Accurate estimates of the risks associated with different types of experiments are difficult to obtain because of our ignorance of the probability that the anticipated dangers will manifest themselves. Nevertheless, experiments involving the construction and propagation of recombinant DNA molecules using DNA's from (i) prokaryotes, bacteriophages, and other plasmids; (ii) animal viruses; and (iii) eukaryotes have been characterized as minimal, low, moderate, and high risks to guide investigators in their choice of the appropriate containment. These designations should be viewed as interim assignments which will need to be revised upward or downward in the light of future experience.

The recombinant DNA molecules themselves, as distinct from cells carrying them, may be infectious to bacteria or higher organisms. DNA preparations from these experiments, particularly in large quantities, should be chemically inactivated before disposal.

1. Prokaryotes, bacteriophages, and bacterial plasmids. Where the construction of recombinant DNA molecules and their propagation involve prokaryotic agents that are known to exchange genetic information naturally, the experiments can be performed in minimal-risk containment facilities. Where such experiments pose a potential hazard, more stringent containment may be warranted.

Experiments involving the creation and propagation of recombinant DNA molecules from DNA's of species that ordinarily do not exchange genetic information generate novel biotypes. Because such experiments may pose biohazards greater than those associated with the original organisms, they should be performed, at least, in low-risk containment facilities. If the experiments involve either pathogenic organisms or genetic determinants that may increase the pathogenicity of the recipient species, or if the transferred DNA can confer upon the recipient organisms new metabolic activities not native to these species and thereby modify its relationship with the environment, then moderate- or high-risk containment should be used.

Experiments extending the range of resistance of established human pathogens to therapeutically useful antibiotics or disinfectants should be undertaken only under moderate- or high-risk containment, depending upon the virulence of the organism involved.

2. Animal viruses. Experiments involving linkage of viral genomes or genome segments to prokaryotic vectors and their propagation in prokaryotic cells should be performed only with vector-host systems having demonstrably restricted growth capabilities outside the laboratory and with moderate-risk containment facilities. Rigorously purified and characterized segments of nononcogenic viral genomes or of the demonstrably nontransforming regions of oncogenic viral DNA's can be attached to presently existing vectors and propagated in moderate-risk containment facilities; as safer vector-host systems become available such experiments may be performed in low-risk facilities.

Experiments designed to introduce or propagate DNA from nonviral or other low-risk agents in animal cells should use only low-risk animal DNA's as vectors (for example, viral or mitochondrial), and manipulations should be confined to moderate-risk containment facilities.

3. Eukaryotes. The risks associated with joining random fragments of eukaryote DNA to prokaryotic DNA vectors and the propagation of these recombinant DNA's in prokaryotic hosts are the most difficult to assess.

A priori, the DNA from warm-blooded vertebrates is more likely to contain cryptic viral genomes potentially pathogenic for man than is the DNA from other eukaryotes. Consequently, attempts to clone segments of DNA from such animals and particularly primate genomes should be performed only with vector-host systems having demonstrably restricted growth capabilities outside the laboratory and in a moderate-risk containment facility. Until cloned segments of warm blooded vertebrate DNA are completely characterized, they should continue to be maintained in the most restricted vector-host system in moderate-risk containment laboratories; when such cloned segments are characterized, they may be propagated as suggested above for purified segments of virus genomes.

Unless the organism makes a product known to be dangerous (for example, a toxin or virus), recombinant DNA's from cold-blooded vertebrates and all other lower eukaryotes can be constructed and propagated with the safest vector-host system available in low-risk containment facilities.

Purified DNA from any source that performs known functions and can be judged to be nontoxic may be cloned with currently available vectors in low-risk containment facilities. (Toxic here includes potentially oncogenic products or substances that might perturb normal metabolism if produced in an animal or plant by a resident microorganism.)

4. Experiments to be deferred. There are feasible experiments which present such serious dangers that their performance should not be undertaken at this time with the currently available vector-host systems and the presently available containment

capability. These include the cloning of recombinant DNA's derived from highly pathogenic organisms (that is, Class III, IV, and V etiologic agents as classified by the U.S. Department of Health, Education and Welfare), DNA containing toxin genes, and large-scale experiments (more than 10 liters of culture) using recombinant DNA's that are able to make products potentially harmful to man, animals, or plants.

IV. Implementation

In many countries steps are already being taken by national bodies to formulate codes of practice for the conduct of experiments with known or potential biohazard (2). Until these are established, we urge individual scientists to use the proposals in this document as a guide. In addition, there are some recommendations which could be immediately and directly implemented by the scientific community.

A. Development of Safer Vectors and Hosts. An important and encouraging accomplishment of the meeting was the realization that special bacteria and vectors, which have a restricted capacity to multiply outside the laboratory, can be constructed genetically, and that the use of these organisms could enhance the safety of recombinant DNA experiments by many orders of magnitude. Experiments along these lines are presently in progress and, in the near future, variants of bacteriophage, nontransmissible plasmids, and special strains of *Escherichia coli* will become available. All of these vectors could reduce the potential biohazards by very large factors and improve the methodology as well. Other vector-host systems, particularly modified strains of *Bacillus subtilis* and their relevant bacteriophages and plasmids, may be found for eukaryotic hosts such as yeast and readily cultured plant and animal cells. There is likely to be a continuous development in this area, and the participants at the meeting agreed that improved vector-host systems which reduce the biohazards of recombinant DNA research will be made freely available to all interested investigators.

B. Laboratory Procedures. It is the clear responsibility of the principal investigator to inform the staff of the laboratory of the potential hazards of such experiments, before they are initiated. Free and open discussion is necessary so that each individual participating in the experiment fully understands the nature of the experiment and any risk that might be involved. All workers must be properly trained in the containment procedures that are designed to control the hazard, including emergency actions in the event of a hazard. It is also recommended that appropriate health surveillance of all personnel, including serological monitoring, be conducted periodically.

C. Education and Reassessment. Research in this area will develop very quickly, and the methods will be applied to many different biological problems. At any given time it is impossible to foresee the entire range of all potential experiments and make judgments on them. Therefore, it is essential to undertake a continuing reassessment of the problems in the light of new scientific knowledge. This could be achieved by a series of annual workshops and meetings, some of which should be at the international level. There should also be courses to train individuals in the relevant methods, since it is likely that the work will be taken up by laboratories which may not have had extensive experience in this area. High priority should also be given to research that could improve and evaluate the containment effectiveness of new and existing vector-host systems.

V. New Knowledge

This document represents our first assessment of the potential biohazards in the light of current knowledge. However, little is known about the survival of laboratory strains of bacteria and bacteriophages in different ecological niches in the outside world. Even less is known about whether recombinant DNA molecules will enhance or depress the survival of their vectors and hosts in nature. These questions are fundamental to the testing of any new organism that may be constructed. Research in this area needs to be undertaken and should be given high priority. In general, however, molecular biologists who may construct DNA recombinant molecules do not undertake these experiments and it will be necessary to facilitate collaborative research between them and groups skilled in the study of bacterial infection or ecological microbiology. Work should also be undertaken which would enable us to monitor the escape or dissemination of cloning vehicles and their hosts.

Nothing is known about the potential infectivity in higher organisms of phages or bacteria containing segments of eukaryotic DNA, and very little is known about the infectivity of the DNA molecules themselves. Genetic transformation of bacteria does occur in animals, suggesting that recombinant DNA molecules can retain their biological potency in this environment with recombinant DNA molecules. It will be necessary to ensure that this work will be planned and carried out; and it will be particularly important to have this information before large-scale applications of the use of recombinant DNA molecules are attempted.

REFERENCES AND NOTES

1. P. Berg et al., *Science,* vol. 185, no. 303 (1974).
2. Advisory Board for the Research Councils, "Report of the Working Party on the Experimental Manipulation of the Genetic Composition of Micro-Organisms, Presented to Parliament by the Command of Her Majesty, January 1975" (Her Majesty's Stationery Office, London, 1975); National Institutes of Health Recombinant DNA Molecule Program Advisory Committee, Bethesda, Maryland.
3. The work of the committee was assisted by the National Academy of Sciences National Research Council Staff; Artemis P. Simopoulos, Executive Secretary, Division of Medical Sciences, Assembly of Life Sciences. Supported by the National Institutes of Health (contract NO1-OD-52103) and the National Science Foundation (grant GBMS75-05293). Requests for reprints should be addressed to: Division of Medical Sciences, Assembly of Life Sciences, National Academy of Sciences, 2101 Constitution Avenue, NW, Washington, D.C. 20418.

Appendix VI
Excerpts from Private Letters and from Addresses *

Contents pertain to:

1) Watson to Sydney Brenner, July 26, 1954 (letter).
2) Letter from Paul Zamecnik describing first clash of two views: Watson concerned with flux of information, giving no explicit place to enzymes; Zamecnik concerned with flow of energy, with enzymes, apparently unaware of nucleic acids.
3) Crick in his extraordinary paper, "Degenerate Templates and the Adaptor Hypothesis."
4) Crick's nine-page handwritten letter to Brenner, July 1955.
5) Crick to Brenner, October 21st.
6) Zamecnik in his notebook, November 1955.
7) Crick to Brenner, February 13, 1956, about a talk Leslie Orgel gave in Cambridge on the coding problem; represents the beginning of, as Judson describes it, "the comma-free code—an idea of Crick's that was the most elegant biological theory ever to be proposed and proved wrong."
8) Maurice Wilkins concluding the paper he delivered, in review of then existing evidence about the structure of DNA in fibres and sperm heads, at a symposium on nucleic acids, organized by the Biochemical Society, London, February, 1956.
9) Crick on the adaptor hypothesis.
10) Crick in a paper delivered at a London symposium on the nature of viruses, organized by the Ciba Foundation, March, 1956.
11) Crick to Jacques Monod, 1962.

---

* Exerpts are derived from those prepared by Judson, Horace Freedlander, *The Eighth Day of Creation*. Reprinted by permission of SIMON & SCHUSTER, a Division of Gulf & Western Corporation.

12) Crick to Brenner, April 30, 1956.
13) The scheme for a comma-free code as worked out by Crick, Orgel, and the theoretical physicist John Griffith.
14) Crick's 16 page letter to Brenner, July 17, 1956.
15) Crick's paper "On Protein Synthesis" which "permanently altered the logic of biology [for] even the errors make the kind of sense they do because of the principles that articulate the whole" (Judson).
16) Crick's report to NATURE, 1970, clarifying his idea of the central dogma of molecular biology.
17) Crick's report on "The Present Position of the Coding Problem" delivered at the Brookhaven National Laboratory, Long Island, New York, June 1, 1959.
18) Crick to Brenner, 1956.

1) In the summer of 1954 we inquired of our scientific neighbor, Paul Doty (a biochemist at Harvard), as to how the RNA of the ribonucleoprotein particle could serve to order the activated amino acids. Dr. Doty mentioned that a Dr. James Watson was visiting him, who with a colleague, Dr. Crick, had recently proposed a model structure for DNA. . . . He would send R. Watson over to visit me and I could ask him that question. I looked at Dr. Watson's young face above his white Irish turtle neck sweater, then at his wire model of the double-stranded DNA, and inquired how the message of the DNA made its way into the sequence of protein. Unfortunately, it seemed to me privately, the bases were facing in, rather than out, as on the biologically appealing earlier Pauling model. Was protein made directly on a DNA template? No answer to this question either, although it seemed likely. How did this complicated double helix unwind? There was a gulf between DNA and protein synthesis, Dr. Watson agreed with a diffident smile, as we parted and he took off on a vacation to look at birds.

2) The Physiology Course is frightfully intense and I have never worked so hard in my life. I shall be in Woods Hole till the time of the phage meeting (Aug 25) when I shall drive down to C. S. H. I would be delighted to see you here. Any time in August is convenient provided you let me know a few days in advance. Francis Crick will also be here as will Gamow. Why not come for a week or so.

3) This difficulty brings us face-to-face with one of the most puzzling features of the DNA structure—the fact that it is *non-polar,* due to the dyads at the side; or put another way, that one chain runs up while the other runs down. It is true that this only applies to the backbone, and not to the base sequence, as Delbruck has emphasized to me in correspondence. This may imply that a base sequence read one way makes sense, and read the other way makes nonsense. Another difficulty is that the assumptions made about which diamonds are equivalent are not very plausible. . . . [Gamow's idea] would not be unreasonable if the amino acid could fit on to the template from either side, into cavities which were in a plane, but the structure certainly doesn't look like that. The bonds seem mainly to stick out perpendicular to the axis, and the template is

really a surface with knobs on, and presents a radically different aspect on its two sides. . . .

What, then are the novel and useful features of Gamow's ideas? It is obviously not the idea of amino acids fitting on to nucleic acids, nor the idea of the bases sequence of the nucleic acids carrying the information. To my mind Gamow has introduced three ideas of importance:

(1) In Gamow's scheme several different base sequences can code for one amino acid. . . . This "degeneracy" seems to be a new idea, and, as discussed later, we can generalise it.
(2) Gamow boldly assumed that code would be of the overlapping type. . . . Watson and I, thinking mainly about coding by hypothetical RNA structures rather than by DNA, did not seriously consider this type of coding.
(3) Gamow's scheme is essentially abstract. It originally paid lipservice to structural considerations, but the position was soon reached when "coding" was looked upon as a problem in itself, independent as far as possible of how things might fit together.

. . . Such an approach, though at first sight unncessarily abstract, is important. Finally it is obvious to all of us that without our President the whole problem would have been neglected and few of us would have tried to do anything about it.

As regards chemical character, I want to consider not only the DNA structure, but also any conceivable form of RNA structure. Now what I find profoundly disturbing is that I cannot conceive of any structure (for either nucleic acid) acting as a direct template for amino acids, or at least as a specific template. In other words, if one considers the physical-chemical nature of the amino acid side chains we do not find complementary features on the nucleic acid. Where are the knobby hydrophobic [water-repelling] surfaces to distinguish valine from Leucine and isoleucine? Where are the charged groups, in specific positions, to go with the acidic and basic amino acids? It is true that a "Teller" scheme, in which the amino acids already condensed act effectively as part of the template, might be a little easier, but a study of [known] sequences from this point of view is not encouraging.

I don't think that anybody looking at DNA or RNA would think of them as templates for amino acids were it not for other, indirect evidence.

What the DNA structure does show (and probably RNA will do the same) is a specific pattern of hydrogen bonds, and very little else. It seems to me, therefore, that we should widen our thinking to embrace this obvious fact. Two schemes suggest themselves. In the first small molecules (phospholipides? ions chelated on guanine? [he was imagining small structures that might attach]) could condense on the nucleic acid and pad it suitably, and the resulting combination would form the template. I shall not discuss this further here. In the second, each amino acid would combine chemically, at a special enzyme, with a small molecule which, having a specific hydrogen-bonding surface, would combine specifically with the nucleic acid template. This combination would also supply the energy necessary for polymerisation. In its simplest form there would be 20 different kinds of adaptor molecule, one for each amino acid, and 20 different enzymes to join the amino acids to their adaptors. Sydney Brenner, with whom I have discussed this idea, calls this the "adaptor hypothesis", since each amino acid is fitted with an adaptor to go on to the template.

The usual argument presented against this latter scheme is that no such small molecules have been found, but this objection cannot stand. For suppose, as is probable, that the small adaptor molecules are in short supply. Then consider the experiment in which all amino acids except one (say leucine), is supplied to an organism, so that protein synthesis stops. Why do not the intermediaries—the (amino acid + adaptor) molecules—accumulate? Simply because there is very little of them, and no more amino acids can combine with these adaptors until the amino acids, to which they are at that moment attached, have been made into proteins, thus releasing the adaptor molecule. Thus under these conditions free amino acids accumulate, not amino acids-plus-adaptor molecules. . . .

In any case it seems unlikely that totally free amino acids actually go on to the template, because a free energy supply is necessary, especially when one bears in mind the entropy contribution [that is, as he explains in a moment, the cost, in energy drawn from elsewhere, of preventing random errors] needed to assemble the amino acids in the correct order. Free energy must be supplied to prevent mistakes in sequence being made too frequently.

The adaptor hypothesis implies that the actual set of twenty amino acids found in proteins is due either to a historical accident or to biological selection at an extremely primitive stage. This is not impossible, since once the twenty had been fixed it would be very difficult to make a change without altering every protein in the organism, a change which would almost certainly be lethal. It is perhaps surprising that an occasional virus has not done this, but even there a number of steps would be required. . . .

It is also conceivable that there is more than one adaptor for one amino acid, and the number 20 may be simply an accident (in any case we need a code for "end chain", so perhaps 21 would be more reasonable).

I have tactily dealt with DNA throughtout, but the arguments would carry over to some types of RNA structure. If it turns out that DNA, in the double-helix form, does not act directly as a template for protein synthesis, but that RNA does, many more families of codes are of course possible. (Incidentally the protein sequences we use to test our theories—insulin, for example—are probably RNA-made proteins. Perhaps a special class of DNA-made proteins exists, almost always in small quantities (and thus normally overlooked), except perhaps where there are giant chromosomes.). . . . In particular base pairing may be absent in RNA or take a radically different form. . . . Without a structure for RNA one can only guess.

Altogether the position is rather discouraging. Where on the one hand the adaptor hypothesis allows one to construct, in theory, codes of bewildering variety, which are very difficult to reject in bulk, the actual sequence data, on the other hand, gives hardly any hint of regularity, or connectedness, and suggests that all, or almost all sequences may be allowed. In the comparative isolation of Cambridge I must confess that there are times when I have no stomach for decoding.

(4) First, about coding. As far as I can see things are still as negative as ever (Incidentally the quotation—Kai Kā'ūs—was genuine!)* Gamow has conceived some enthusiasm for the "combination code", mainly because it gives different frequencies for different amino acids. But its so hard to think of a structural basis, and the supporting evidence is so weak, that I cannot really take it seriously. I have just received, but only read quickly, the final draft of Gamow, Rich & Ycas. The basic difficulty, as they point out, is that so far there is nothing striking to explain. My own view is that "coding", etc, should be put on the shelf for a bit. I think it may help to complete the picture when more experimental facts are known, but I don't think it is likely to go much further on its own.

Re DNA structure. Nothing new, beyond what you have read in the Wilkins & Co letter in Nature recently. . . . I'm told Gamow has a paper in NAS [*Proceedings of the National Academy of Sciences*] . . . which amounts to paranemic coiling. Jim says that Gamow sent it to Delbruck, who advised against publication. . . .

The Ochoa story, which I expect has reached you. An enzyme preparation which polymerises such things as ADP, to give, with the loss of one phosphate, a polymer i.e., "RNA". Will do it for other bases, and will produce copolymers. Jim says that Alex has got the X-ray powder diagram [that is, diffraction picture, but not from a crystalline form and therefore diffuse]. . . . , and it looks similar to RNA proper!! We have hopes of having Alex here before long (Jim arrived a few days ago. He has a job at Harvard, Biology Dept., but has leave of absence, and hopes to stay here for 6 months) so we may have the story in more detail.

(5) Your letters arrived the day after mine went off. I was delighted to hear that you were prepared, if necessary, to work in a cupboard!

(6) We were . . . anxious to remove all doubts that (a) the amino acid was covalently bound, (b) that the aminoacyl nucleotide [the two components as bound together] was a direct line intermediate in the pathway from free amino acid to peptide, and not a side path for storage of activated amino acids, and (c) that the entire complex polynucleotide—and not a much smaller triplet or oligomer [short chain] which adventitiously aggregated with this larger type of RNA—was the active intermediate.

(7) This has started us off again, but nothing really exciting. We are looking into the problem: if it is three non-overlapping bases per amino acid, how does one get over the difficulty that there is no comma to show how the bases are grouped in threes? Will let you know how this turns out.

(8) It is only in the last year or two that any convincing demonstration has been made that protein synthesis involves nucleic acids, and that a structure of one of the nucleic acids (and possibly not the whole of it) has been established. . . . It is, therefore, not surprising that little of a constructive nature can be said at present of the implications

* "Is there anyone so utterly lost as he that seeks a way where there is no way," Kai Kā'ūs ibn Iskander ibn Qābūs, Amir of Dailam on the Caspian, a minor medieval Persian writer.

of nucleic acid structure with respect to protein synthesis. . . . It seems unlikely that theoretical approaches can tell us much about protein synthesis until a great deal more experimental work has been done.

9) What a particular sequence of bases does provide is a highly specific pattern of sites for hydrogen bonding. It is thus reasonable to expect that whatever is absorbed onto the nucleic acids in a specific manner will be capable of forming a definite spatial pattern of hydrogen bonds. This can be envisaged in many ways. One possibility is that each amino acid is combined with its special "adaptor". These might be any sort of small molecule . . . but an obvious class would consist of molecules based on di-, tri-, or tetra-nucleotides. The fact that such molecules have not yet been discovered is not a strong argument against this sort of scheme, since they may be somewhat unstable. . . . The combination between each amino acid and its own specific adaptor could be made by a special enzyme, or enzymes, which would give a better opportunity for preventing "mistakes", such as valine occurring in an isoleucine position.

This idea would thus be in line both with many typical biochemical systems and with the physical-chemical nature of the nucleic acids. The essential point is that each amino acid should have its own specific adaptor, capable of forming hydrogen bonds.

0) The information required to synthesize the virus protein is contained in the RNA. As there is only a limited amount of RNA it can only carry a limited amount of information. Thus the protein molecules of the virus can only be of limited size. Rough numerical estimates show that this amount, used once, is not enough to produce a shell to cover the RNA. Thus the coat must be built up of identical subunits.

1) I am always surprised that people do not refer more to our ideas on virus structure. The idea of subuntis in spherical viruses had been suggested before, but it had made no impression. All the modern X-ray work on viruses came about because of our influence, including Rosalind Franklin's excellent work on TMV, which was based on Jim's evidence that the structure was helical. . . . The basic reason for subunits—that the information the RNA could carry is limited—was also quite original. At the time it even seemed daring!

12) I leave tomorrow for 3 months in the States. . . . I think when I return in August I should start looking into the business of a school for Jonathan (Brenner's stepson). . . . Re coding. Leslie, John Griffith & I have deduced the magic 20, using a code having 3 bases to 1 amino acid. (Did I tell you about this). A RNA tie-club account of this will reach you shortly. It's clever, but I can't make up my mind as to its correctness. . . . Jim, Alex and I have a structure for Ochoa's poly A. . . . Looks promising. . . . RNA may be several of these twisted together, but the point escapes us. Our latest idea is that microsomes may have spherical symmetry & contain protein subunits. . . . . Vernon thinks we have two chicks with different lysozymes. I am doubtful.

13) Using the metaphors of coding, we say that some of the 64 triplets make sense and some make nonsense. We further assume that all possible sequences of the amino acids

may occur (that is, can be coded) and that at every point in the string of letters one can only read "sense" in the correct way. . . . In other words, any two triplets which make sense can be put side by side, and yet the overlapping triplets so formed must always be nonsense.

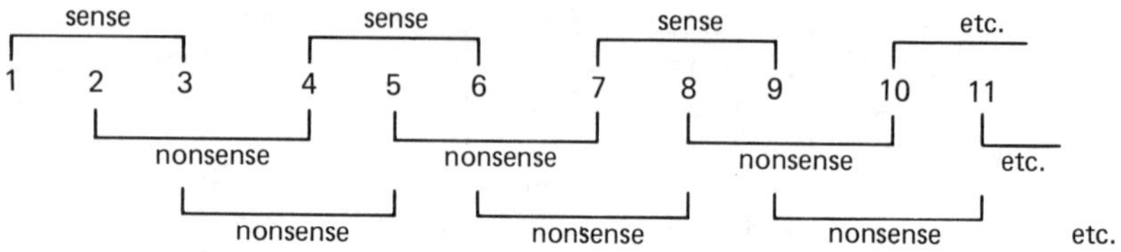

We can write down a construction which obeys all our rules and yet codes 20 amino acids. One possible solution is

```
                A          A       A   A
         A    A   B                    B
   A B           C           B D C
         B    B   C           C    D
```

where A B $\genfrac{}{}{0pt}{}{A}{B}$ means ABA and ABB, etc. It is easy to see, by systematic enumeration, that one can place any two triplets of this set next to each other without producing overlapping triplets that belong to the set.

To fix ideas, we shall describe a simple model to illustrate the advantages of such a code. Imagine that a single chain of RNA, held in a regular configuration, is the template. Let the intermediates in protein synthesis be 20 distinct molecules, each consisting of a trinucleotide chemically attached to one amino acid. The bases of each trinucleotide are chosen according to the code given above. Let these intermediate molecules combine, by hydrogen bonding between bases, with the RNA template. . . . Now imagine that such an amino acid-trinucleotide were to diffuse into an incorrect place on the template, such that two of its bases were hydrogen-bonded, though not the third. We postulate that this incomplete attachment will only retain the intermediate for a very brief time (for example, less than 1 millisecond [thousandth of a second] before the latter breaks loose and diffuses elsewhere. However, when it eventually diffuses to the correct place, it will be held by hydrogen bonds to all three bases and will thus be retained, on the average, for a much longer time (say, seconds or minutes). Now the code we have described insures that this more lengthy attachment can occur only at the points where the intermediate is needed. If one of the 200 intermediates could stay for a long time on one of the false positions, it would effectively block the two positions it was straddling. . . . Our code makes this impossible.

(14) Gunther's idea is that DNA makes RNA . . . which then pairs with another similar strand of RNA. A new DNA chain is then made on this & the RNA chain is destroyed

(to explain the distribution of mutants) and the DNA (old & new) then makes more RNA, and so on.

Personally I don't like it, for a number of reasons. But Kornberg's system (see later) might decide. . . . It really looks as if he may have specific DNA replication in the test-tube. He works with "enzyme systems" as from Coli. He finds:

1. He needs a primer. This primer is intact DNA. Even a very light digestion with DNAse destroys its priming ability. Good DNA from several sources works equally well, except for phage (T2?) which works only half as well (this may not be significant).
2. He gets net synthesis: usually about 10 to 50% increase in the DNA used as a primer. The exact amount is not yet significant. . . .
3. The precursors are the (deoxy) tri-phosphates. . . . All four are required at once. Omit any one (or two) and the rate of incorporation drops from 50 to 0! (They had a lot of trouble with the original system, as they had varying amounts of the precursors there as contaminants.)
4. RNAse (the enzyme) added both before and during incubation makes no difference (if they really have specific replication this disposes of Gunther).

They have not yet checked if the base-ratios are copied. . . . Everybody was greatly impressed by the great competence of Kornberg. Its my belief that he really has DNA replication and that he will prove it within a very short time.

15) It is at first sight paradoxical that it is probably easier for an organism to produce a new protein than to produce a new small molecule, since to produce a new small molecule one or more new proteins will be required in any case to catalyse the reactions.

I shall . . . argue that the main function of the genetic material is to control (not necessarily directly) the synthesis of proteins. There is a little direct evidence to support this, but to my mind the psychological drive behind this hypothesis is at the moment independent of such evidence. Once the central and unique role of proteins is admitted there seems little point in genes doing anything else.

What can we guess about the structure of the microsomal particle? On our assumptions the protein component of the particles can have no significant role in determining the amino acid sequence of the proteins which the particles are producing. We therefore assume that their main function is a structural one, though the possibility of some enzyme activity is not excluded. The simplest model then becomes one in which each particle is made of the same protein, or proteins, as every other one in the cell, and has the same basic arrangement of the RNA, but that different particles have, in general, different base-sequences in their RNA, and therefore produce different proteins. This is exactly the type of structure found in tobacco mosaic virus, where the interaction between RNA and protein does not depend upon the sequence of bases of the RNA. . . . In addition Watson and I have suggested . . . , by analogy with the spherical viruses, that the protein of microsomal particles is probably made of many identical sub-units arranged with cubic symmetry. . . .

It would at least be of some help if the approximate location of the RNA in the microsomal particles could be discovered. Is it on the outside or the inside of the particles, for example, or even both? Is the microsomal particle a rather open structure, like a sponge, and if it is what size of molecule can diffuse in and out of it? Some of these points are now ripe for a direct experimental attack.

It is instructive to compare your own hemoglobin with that of a horse. Both molecules are indistinguishable in size. Both have similar amino acid compositions; similar but not identical. They differ a little electrophoretically, form different crystals, and have slightly different ends to their polypeptide chains. All these facts are compatible with their polypeptide chains having similar amino acid sequences, but with just a few changes here and there.

This "family likeness" between the "same" protein molecules from different species is the rule rather than the exception. It has been found in almost every case in which it has been looked for. One of the best studied examples is that of insulin, by Sanger and his co-workers [Crick cited recent papers], who have worked out the complete amino acid sequences for five different species, only two of which (pig and whale) are the same. Interestingly enough the differences are all located in one small segment of the two chains.

Biologists should realize that before long we shall have a subject which might be called "protein taxonomy"—the study of the amino acid sequences of the proteins of an organism and the comparison of them between species. It can be argued that these sequences are the most delicate expression possible of the phenotype of an organism [that is, of the individual's observable characteristics as determined by its genes] and that vast amounts of evolutionary information may be hidden away within them.

My own thinking (and that of many of my colleagues) is based on two general principles, which I shall call the Sequence Hypothesis and the Central Dogma. The direct evidence for both of them is negligible, but I have found them to be of great help in getting to grips with these very complex problems. I present them here in the hope that others can make similar use of them. Their speculative nature is emphasized by their names. It is an instructive exercise to attempt to build a useful theory without using them. One generally ends in the wilderness.

The Sequence Hypothesis. This has already been referred to a number of times. In its simplest form it assumes that the specificity of a piece of nucleic acid is expressed solely by the sequence of its bases, and that this sequence is a (simple) code for the amino acid sequence of a particular protein.

This hypothesis appears to be rather widely held. Its virtue is that it unites several remarkable pairs of generalizations: the central biochemical importance of proteins and the dominating biological role of genes, and in particular of their nucleic acid; the linearity of protein molecules (considered covalently) and the genetic linearity within the functional gene, as shown by the work of Benzer . . . and Pontecorvo [Crick cited their papers]; the simplicity of the composition of protein molecules and the simplicity of the nucleic acids. Work is actively proceeding in several laboratories, including our own, in an attempt to provide more direct evidence for this hypothesis.

The Central Dogma. This states that once "information" has passed into protein it cannot get out again. In more detail, the transfer of information from nucleic acid to nucleic acid, or from nucleic acid to protein may be possible, but transfer from protein to protein, or from protein to nucleic acid is impossible. Information means here the precise determination of sequence, either of bases in the nucleic acid or of amino acid residues in the protein.

This is by no means universally held . . . but many workers now think along these lines. As far as I know it has not been explicitly stated before.

16) The transfer protein → RNA (and the analogous protein → DNA) would have required (back translation, that is, the transfer from one alphabet to a structurally quite different one. It was realized that forward translation involved very complex machinery. Moreover, it seemed unlikely on general grounds that this machinery could easily work backwards. The only reasonable alternative was that the cell had evolved an entirely separate set of complicated machinery for back translation, and of this there was no trace, and no reason to believe that it might be needed. . . .

In looking back I am struck not only by the brashness which allowed us to venture powerful statements of a very general nature, but also by the rather delicate discrimination used in selecting what statements to make. Time has shown that not everybody appreciated our restraint.

17) This large variation of DNA composition is very unexpected. The abundance of the various amino acids does not, as far as we know, vary much from organism to organism; leucine is always common, methionine usually rather rare. The small variation of RNA composition is exactly what might be expected; the large variation reported for DNA needs some special explanation. . . . Listed below are some possible explanations of this phenomenon, though in my view they all, at the moment, appear unattractive. . . .

1. Only part of the DNA codes protein. . . . The difficulty of this idea is that the nonsense must make up a rather large fraction of the DNA. If, for example, it is assumed that the base composition of the sense is reflected in that of the total RNA of the organism, then organisms showing extreme base ratios must have a minimum of 35% nonsense in their DNA. . . . A possible reason for the fine dispersion of nonsense might be the provision of "commas.". . .

2. The DNA-to-RNA translation mechanism varies. This would allow the RNA-to-protein code to be uniform throughout nature, while permitting the DNA-to-RNA code to vary. This is a formal possibility, but it does not appear at all likely. . . .

3. The code is degenerate. . . .

4. The code is not universal. It used to be argued that the code would be uniform throughout nature (except possibly for certain virulent viruses) because any attempt to change it would necessarily alter many proteins at once and would thus almost certainly be lethal. However, a counterargument has been given (it had been suggested to Crick by Cyrus Levinthal, from MIT, who was in the audience at Brookhaven) . . . that perhaps under certain conditions (for example, when the organism was in a rich envi-

ronment and thus did not require too many enzymes) it might be possible to make a change. . . . It does not seem that this point can be usefully argued.

5. The nucleic acid code has less than four letters. . . .

6. The amino acid composition of the protein varies. Unfortunately a small variation will not do. The organisms with extreme base ratios in their DNA are required to have proteins for which, say, leucine is rare and methionine common. This . . . does not seem very likely.

Finally . . . a few important experimental facts should be mentioned which any theory will have to explain: first, the evidence that the RNA of tobacco mosaic virus controls, at least in part, the amino acid sequence of the protein of the virus; second, the genetic effects of transforming factor, which appears to be pure DNA; and third, the genetic control of at least parts of the amino acid sequence of human hemoglobin.

(18) This must be most of the news—if only in outline. There are so many things I should like to do that I can hardly wait for you to arrive. Is there no chance of your arriving a month or two earlier? (i.e. you might fly ahead of your family so that you could find a house for them (ingenious).) In any case we should start thinking about equipment now. . . . Please forgive me if some of this is a little incoherent.

GLOSSARY

Abiogenesis. Spontaneous generation.

Acellular. Consisting of one cell.

Acquired Characteristic. Controversial theory of variations acquired as responses to environmental influences, and transmittable to offspring.

Adaptation. In evolution, any characteristic of living organisms which, in the environment they inhabit, improves their chances for survival.

Adaptive Radiation. In evolution, adaptation to distinct modes of life. Relates to primitive types of organisms radiating into many forms adapted to running, flying, swimming, burrowing, etc.

Allele(s), Allelomorph(s). Alternate forms of a gene, one of two or more forms. Genes grouped upwards of two are alleles of each other and allelic or allelomorphic to each other: occupying same relative position (locus) on homologous chromosomes. When in same cell undergo pairing at meiosis. Effect, though differently, same set of developmental processes. Mutation can occur one to another.

Allograft. Graft in which donor has genotype different from that of host.

Alternation of Generations. Regarding minor animals, some plants: in life cycle, sexual reproduction regenerationally alternated with a generation having asexual reproduction.

Amino Acids. Building blocks of proteins, which are made each from hundreds of thousands of amino acid molecules. There are 20 common amino acids. The basic structure is the same in all amino acids, their side groups do, however, differ. All are organic compounds.

Amino Acid Sequence. Linear order of amino acids in peptide or protein.

Angstrom (Å). Unit of length describing atomic dimensions; equal to 10^{-8} cm.

Antagonism. Interference/inhibition in regard to the growth of one kind of organism by another. Occurs through the creation of unfavorable conditions, for instance the production of an antibiotic substance such as penicillin, or by exhaustion of food supply.

Antibiotic. Substance produced in organisms, diffused into its surroundings, and toxic to individuals of certain other species.

Antibody. Protein produced in vertebrates when an antigen normally foreign to body fluids gains access to them.

Antigen. Substance that stimulates production of antibodies.

Artificial Insemination. Technique of injecting semen artificially into female.

Artificial Parthenogenesis. Technique of activating egg development without contact with sperm, with an egg not normally parthenogenetic.

Asexual Reproduction. Reproduction independent of gametes. (Does not include parthenogenesis.)

Autoimmunity. A diseased state in which immunity acts against the animal in which it operates.

Autosomes. Chromosomes other than sex-chromosomes.

Backbone. (1) Atoms of a polymer that are common to all of its molecules. Sugars and phosphates form a backbone in RNA. (2) Spine. (3) Spiral.

Bacterial Viruses. Viruses that multiply in bacteria. *See* Bacteriophages.

Bactericidal. Lethal to bacteria.

Bacteriology. Study of bacteria.

Bacteriophages, Phages. Viruses that parasitize bacteria.

Bacteriostatic. Inhibiting growth of bacteria in such a way that it does not result in their death.

Base Analogs. Purines and pyramidines differing slightly in structure from "normal" nitrogenous bases. Analogs are sometimes in certain cases incorporated into nucleic acids, replacing the normal constituent.

Base Pairing. Two strands of nucleic acid, DNA or RNA, associating laterally by bonds between certain pairs of bases in their nucleotides provides an example of base pairing. Base pairs consist of adenine, thymine (or uracil), guanine, and cytosine in the nucleic acid double helix of DNA. Pairing rules require adenine to pair with thymine (or uracil), and guanine to pair with cytosine.

Base Ratio. The amount of adenine (A) plus thymine (T) to that of guanine (G) plus cytosine (C) in DNA varies widely and determines the base ratio. The amount of A and T equals that of G and C as paired in DNA.

Biological Clocks. Mechanisms allowing genes to be expressed at periodic intervals.

Biological Control. Artificial pest and parasite control using other organisms, such as control of mosquitoes by fishes or by insectivorous plants feeding on larvae.

Biometry. Mathematics as applied to biology; especially statistical analysis of resemblances and differences between groups of related organisms.

Biophysics. Physics as applied to biology.

Biotype. Naturally grouped individuals of similar genotypes.

Breakage and Reunion. Classical crossing-over by physical breakage and "crossways" reunion. Occurs to completed chromatids during meiosis. As a model of crossing-over, applicable in at least one instance on molecular level: occurring before crossing-over between phage-DNA molecules.

Cell Body (Perikaryon). Cytoplasm with contained nucleus; gives rise to branches of the nerve cell.

Cell Cycle. Cyclic changes in replication. The sequential events occur in the period between mitotic divisions.

Cell Differentiation. Process of cellular specialization of structure and function.

Cell Divion. Division of cell, including cytoplasm and nucleus (usually divided by mitosis, occasionally by amitosis), into two complete cells.

Cell-Free Extract. Fluid containing most of the soluble molecules of a cell. The extract is made by breaking cells open and disposing of remaining whole cells.

Cell Fusion. Formation of a single hybrid cell with the nuclei and cytoplasm of different cells. Cell fusion is often induced by treatment of mixed cell culture with killed Sendai virus.

Cell Lineage. Developmental history: descent of cell division of later cells from earlier cells.

Cell Theory. Theory initiated by Schleiden and Schwann, 1838–1839: Cells and their products constitute all animals and plants; growth and reproduction are fundamentally due to cell division.

Cellular Affinity. Tendency of cells to adhere to cells of same type only. This property is lost in cancer cells.

Central Dogma Basic relationship between DNA (template for its own duplication and for synthesis by RNA), RNA (template for protein synthesis), and protein.

Chimera (Chimaera). Organism whose tissues are of two or more genetically different kinds. Can occur especially as a result of mutation, abnormal distribution of chromosomes, or as a result of natural or artificial grafting of two different plants or animals, whose cells become associated, produce mixed characters: Such a plant is often called a *graft hybrid.*

Chromatid. Duplicated chromosome, its two daughter strands still joined by a single centromere.

Chromatid. In chromosomes: nucleoprotein; strains strongly with basic dyes.

Chromosome. In plants and animals: Thread-shaped body that consists largely of DNA and protein.

Cleavage (Segmentation). Repeated subdivision of zygote cytoplasm; accompanies corresponding nuclear mitosis, which follows fertilization.

Cleavage Divisions. Repeated mitotic divisions of fertilized egg until the stage when regions of egg shift relative to one another.

Cleaving. Part of usual process in preparing tissues for microscopical examination.

Clone. (1) The descendant produced vegetatively or by apomixis from a single plant; asexually or by parthenogenesis from a single animal; or by division from a single cell. The members of a clone are of the same genetic constitution, except insofar as mutation occurs among them. (2) A group of cells all descended from a single common ancestor. (3) An offspring genetically identical to the single parent from which it results; produced asexually, by parthenogenesis, or by the genetic engineering technique of cloning.

Codominant. *See* Dominant.

Codon. A sequence of three adjacent nucleotides in RNA or DNA that code an amino acid.

Collagen. Fiber: fibrous protein (found in bone).

Complement. Group of substances normally present in the blood of vertebrates that perform an essential role in the lethal action (lysis) of antibodies on bacteria and other cells.

Complementary Base Sequences. Related (by base-pairing roles) polynucleotide sequences.

Complementary Structures. Of two structures: each of the structures defines the other. The two strands of a DNA helix are complementary structures.

Complementation. Intercistron complementation: Effect of two homologous chromosomes, which are mutant and defective, in different cistrons. Complementation test: Introduction of two mutant chromosomes or sections of thereof, into the same cell; research purpose is that of seeing whether or not their respective mutations occurred in thc same gene.

Conjugation. Union of gametes, which may or may not be free-swimming, especially in isogamy. Union of two cells in certain bacteria, *Escherichia* and related genera. Allowing passage of generic material from one, the donor cell, to the other, the recipient cell.

Contact Inhibition. Temporary cessation of cell movement, often observed when freely growing cells of (derived from) a multicellular organism come into physical contact with each other.

Cross. Act or product of cross-fertilization.

Cross-Fertilization. Fusion of male and female gametes from different individuals of the same species.

Crossing-Over. The exchange of genetic material between homologous chromosomes.

Cross-Pollination. Transference of pollen from stamens of flower of one plant to the stigma of a flower of another plant of the same species.

Cyclic AMP. Adenosine monophosate with phosphate group. Active in regulation of gene expression in both bacterial and eukaryotic cells.

Cytology. Study of cells, somewhat limited to specific cellular constituents.

Cytoplasm. The entire protoplasm of a cell except for the nucleus.

Cytoplasmic Inheritance. Non-Mendelian (extrachromosomal) inheritance. Is produced by self-propagating cytoplasmic entities.

Cytotoxic. That which acts as a poison to cells.

Darwinism. Evolution by natural selection (Charles Darwin).

Daughter Cells (D. Nuclei). Two cells (or nuclei) resulting from division of single cell; mitosis is usually responsible.

Dedifferentiation. *See* Differentiation.

"Degenerate" Code, "Degenerate Codons." Upwards of two codons describing the same amino acid.

Deoxyribonucleic Acid (Desoxyribonucleic Acid). DNA; polymer of deoxyribonucleoticles. The genetic material of cells.

Determinant. Antigenic: That part of antigen molecule responsible for forming a specific bond with a corresponding antibody molecule. Genetic: Factor or transmitting inheritance: a gene or a plasmagene.

Differentiation. Process of change in cells resulting in appearance of the structures and functions characterizing various kinds of cells, tissue, or organs in a body. Must occur at relatively stable phase of body's life history. Before: undifferentiated. After: differentiated. Reverse change: dedifferentiation.

Diploid. Chromosomes, except sex chromosomes, in pairs ($2N$): each pair is homologous; twice the haploid number is present.

DNA. See: Deoxyribonucleic Acid.

Dominant. Gene producing the same characteristics when it is heterozygous to a specific allele as it does when homozygous.

Double Helix. Basic spiraling structure of DNA.

Double Recessive. Individual homozygous for a particular recessive gene. Offspring show dominant phenotype. May also form heterozygous individual—half dominant, half recessive.

Drosophila. Fruit or banana-fly.

E. coli. See Escherichia coli.

Embryo. In botony and zoology, new individual state of formative development.

Embryology. Study of development of embryos.

Enrichment Culture. Technique that allows selection and isolation of microorganisms having growth characteristics desired by the investigator. Selection is from a natural, mixed population. Used widely in bacteriology.

Enzyme. A protein molecule catalyzing chemical reactions; increases reactivity of specific substance or substances in metabolism.

Escherichia coli. Bacterium used extensively in genetic engineering studies.

Eubacteria. The largest and most diverse group of bacteria.

Eucaryote (Eukaryote). Cells or organism in which nucleus is separated from cytoplasm by nuclear membrane; genetic material is borne on chromosomes consisting of DNA and protein. Applies to all organisms except bacteria and blue-green algae.

Eugenics. Study of improving humanity by altering its genetic composition.

Expressivity. Intensity of gene effect as realized in phenotype. Both genotype and environment are influential to varying degrees.

Extracellular Material. The material within an organism but not within its cells.

F_1 (First Filial Generation). Offspring derived from crossing plants or animals of parental generation, P_1.

Fixation. Initial step, requiring cell death, in making permanent preparations for microscopic study of organisms or tissues.

Gamete. Germ cell, which is reproductive cell. Two gametes are fused in fertilization: female gamete is ovum; male gamete is spermatozoon. Usually haploid; some gametes in animals and plants are, however, diploid and can develop new individuals through parthenogenesis (without fertilization).

Gene Exchange. In breeding populations: fertile offspring which possess new combinations of parental genes.

Gene Flow. Gene movement resulting from mating and gene exchange. Occurs within populations.

Gene Frequency. In populations: frequency with which a kind of gene occurs in relation to the frequency of its alleles.

Gene Pool. Collective: all the genes of a population or species.

Generation Time. In a population of cells: mean time between the division of a cell and the divisions of its daughter cells.

Genetic Code. The system of correspondence between nucleotide sequence and amino acid sequence, the latter determined for proteins by the sequence of nucleotides in Messenger RNA, this depending on their sequence in DNA.

Genetic Drift. Change in the genetic constitution of a population. Occurs not from natural selection but by chance. Is virtually exclusive to small, isolated populations.

Genome (Genom). In a species: set of chromosomes in each nucleus.

Genotype. Genetic constitution (versus physical appearance) of an organism.

Germ Cell. Gamete.

Germ Layer. In an embryo: one of the main layers (groups of cells) distinguishable during and immediately after gastrulation.

Germ Plasm. A kind of protoplasm; Weismann suggested its transmission, largely unchanged, from one generation to the next via germ cells.

Haploid. State in which a single set of unpaired chromosomes is present in a nucleus, as with gametes.

Helix. Spiral structure: usually a single spiraling strand present as a whole or as part of a spiral; exhibits a repeating pattern described by simultaneous rotation and translation. Natural to many regular biological polymers.

Heteromorphic. Concerning alternation of generations, generations being vegetatively dissimilar.

Heterozygous. Having two different allelomorphs in the two corresponding loci of a pair of chromosomes.

Hirsute. Of a person, usually exhibiting excessive hairiness, bearing a superficial visual resemblance, therefore, to animals and not infrequently referred to as a wild man or strange beast. Reputed in China and Siberia especially. Such individuals are, some contend, atavisms, expressing the genetic constitution of an early human species. Another theory is that they are direct descendants of the human "ancestor," the ape *Gigantopithecus.* Genetics holds that abnormal characteristics result in such cases from, probably, random recombinations of recessive traits.

Histone. Class of basic proteins unusually rich in basic amino acids, such as lysine; may be concerned with shutting off gene activity; histones occur where the DNA is specifically complexed with protamines (also basic proteins).

Homeostasis. State of maintained constancy of internal environment.

Homologous Chromosomes. Chromosomes pairing at meiosis; share same morphology; contain genes governing same characteristics.

Homozygous. Having identical genes in the two corresponding loci of paired chromosomes. The homozygous organism is a homzygote for that locus.

Hormone. Organic chemical compound, often a small polypeptide, stimulating functional cellular activity; may, for instance, stimulate adenylcyclase in cell membrane to produce cyclic AMP.

Host. Cell whose metabolism is used for growing, reproducing a foreign substance such as a virus.

Hybrid. Plant or animal which results from a cross between parents genetically unlike each other.

Hydrogen Bond. Weak attractive force. Occurs between an electronegative atom and a hydrogen atom, the latter covalently linked to a second electronegative atom.

Hypostasis. An interaction between nonallelic genes. A hypostatic gene will not be expressed in the presence of another hypostatic gene.

Immunogenic. Having the ability to cause the formation of an antibody.

Inbreeding. Mating of closely related individuals.

Inducible Enzyme. Adaptive enzyme; production rate is increasable by presence of inducers in the cell.

Intercellular. Occurring between cells.

Interphase. State of a cell when it is not undergoing mitosis.

Intracellular. Within a cell.

Intraspecific. Among individuals of the same species. Natural selection, for instance, may occur for one individual at the expense of another kind of the same species.

Introgressive Hybridization. Genes of one species infiltrating the genotype of another species.

In Vitro. (Latin "in glass") Isolated from the organism as a whole in its natural environment. Pertains to experiments performed in a cell-free system; may be modified as a term to include cells grown from multicellular organisms under tissue-culture conditions.

In Vivo. (Latin "in life") In the living organism. Pertains to experiments in which organism remains intact whether at the level of the cell or the whole organism.

Isolating Mechanism. Mechanism that restricts exchanges of genes between populations; it diminishes crossbreeding. The evolutionary divergence of populations from each other requires the functioning of this mechanism.

Karyokineses. Mitoses.

Kineses. Concerning locomotion: movement of a cell or organism in response to a particular stimulus.

Lethal Gene. Gene that kills the individual which bears it. A dominant lethal gene results in heterozygous death. A recessive lethal gene results in homozygous death.

Linkage. Association of at least two nonallelomorphic genes positioned on the same chromosome and tending to be passed from generation to generation as an inseparable unit.

Locus. In all homologous chromosomes, the place where a particular gene or one of its alleles is positioned.

Macromolecules. Molecules of very many atoms. Their molecular weights range from a few thousand to hundreds of millions of units (e.g. proteins, nucleic acids, polysaccharides).

Matrix. Intercellular substance in which cells are embedded (e.g. cartilege or connective tissue).

Meiosis (Reduction Division). The process of two successive cell divisions starting with the diploid cell.

Metaphase. A stage occurring in both mitosis and meiosis.

Metazoa. Multicellular animals.

Microdissection. Technique of performing operations with mechanically operated instruments on minute objects such as living cells while these are viewed through a microscope.

Mitosis (Karyokinesis). Common process by which a nucleus divides into two. Mitotic recombination refers to crossing-over between homologous chromosomes, a process which occurs during mitosis and leads to segregation of hererozygous alleles.

Multinucleate. Of a cell: containing many nuclei.

Multiple Allelomorphs. A set of more than two genes, all allelomorphs.

Multiple Factors (Multiple Genes). *See* Polygenes.

Mutagenic. Able to cause mutations: physical or chemical agents, including radiation, heat, or alkylating or deamina deaminating agents raising mutation frequency high above the spontaneous background level.

Mutant. A gene having undergone mutation; an organism bearing such a gene; a character due to such a gene.

Nucleic Acid. Nucleotide polymer: long-chain molecule formed from many nucleotides.

Nucleolus. Round, granular structure with a small dense body containing RNA and protein. There may be one or more typically located in the resting nucleus of eucaryotic cells. The nucleolus is usually associated with a specific chromosomal site. It is involved in rRNA synthesis as well as in ribosome formation.

Nucleoproteins. Compounds of nucleic acids and proteins.

Nucleotide. Compound found free in cells and as part of various coenzymes. Nucleotides form the building blocks of nucleic acids.

Nucleus. Body containing chromosomes in eucaryotes.

Oncogenic. Producing cancer.

Ontogeny. Development: its whole course during the life history of an individual.

Oocyte. That cell which undergoes meiosis and forms an ovum.

Operon. Unit of closely linked genes, controlled by an operator and a repressor, that are functionally related as members of one enzyme system. Synthesizes mRNA which is switched on or off by an operator gene, part of the operon.

Orthogenesis. Steady trend of evolution which moves in a given direction and continues over a prolonged period of time.

Pairing. Sideways attachment of two homologous chromosomes before crossing-over.

Parthenogenesis. Development of new individual from ovum, bypassing fertilization; natural to some plants (dandelion) and animals (aphids, where males are absent). Ovum is usually diploid.

Peptide Bond. Covalent bond between two amino acids.

Perikaryon. *See* Cell Body.

pH. Measure of acidity or alkalinity of a solution.

Phage. *See* Bacteriophage.

Phagocyte. A cell that takes surrounding particles into its cytoplasm, an important defense mechanism against bacteria for most Metazoa.

Phenocopy. Change in the normal phenotype that is induced by environmental influences at development; not inherited but mimics gene mutation.

Phenotypye. Set of observable properties basic to an organism; the properties are produced by the genotype with cooperation from the environment.

Plaques. Areas in a confluent cell sheet. The areas are round and clear. Plaques result from killing or the lysis of contiguous cells; this itself results from several cycles of virus growth.

Plasma Membrane. Cell membrane.

Plasmon. Collectively extrachromosomal hereditary factors.

Plastogene. Factor having uninherited (Mendelian fashion) properties of a gene; carried by a plastid.

Pleiotrophic. Of a gene: affecting more than a single characteristic.

Polygenes. Any of those genes that produce but a slight effect on phenotype differences. Systems of polygenes are associated with producing quantitative variations (may also be environmentally produced) of certain characteristics such as stature.

Polypeptide. A polymer (regular subunit monomers covalently bonded) of amino acids. The amino acids are linked together by peptide bonds.

Polyploid. Having at least three times the haploid number of chromosomes (common in plants).

Primordial Germ Cells. In vertebrates, large cells appearing in early embryonic life and migrating to gonads; give rise to gametes.

Procaryote. Unicellular organism such as bacterium. Does not have a nuclear membrane, membrane-bound organelles, or characteristic ribosomes or biochemistry.

Prophage. DNA integrated into host cell and reproduced as part of the host.

Pure Line. Collectively, successive generations of organisms that are homozygous for all genes, variation usually resulting from noninherited modifications produced by the environment; produced also by intensive inbreeding.

Qualitative Inheritance. Expression of a specific character in sharp contrast among individuals of a species (discontinuous variation); depends on a few genes producing a major effect.

Quantitative Inheritance. The usually gradated (degrees from one extreme to another) expression of character differences among individuals of a species (continuous variation); depends on many genes producing small effects.

Recapitulation. In embryonic development: stages repeating the structure of the adult forms of ancestors.

Receptor Site. Area of cell or molecule that combines with an agent such as a hormone molecule or a virus.

Recessive. Of genes: not affecting the phenotype in which recessivity is present, except when individual is homologous.

Recombination. Appearance in offspring of traits that did not appear together in this way in either parent; may occur naturally by independent assortment.

Reduction Division. *See* Meiosis.

Replication. Exact copying of complex molecules, a process necessary during the growth and reproduction of a life form. The template copied is usually the DNA molecule, sometimes RNA.

Resting Cell. Cell or nucleus at interphase.

Ribonucleic (Ribosenucleic) Acid (RNA). Polymer composed of ribonucleotides.

Ribosomes. Tiny (about 200 **9** in diameter) cellular particles, consisting of rRNA and protein, sites of protein synthesis.

Selection Pressure. Measure regarding evolution: extent to which the genetic constitution of a population is altered by natural selection.

Somatic Cells (Soma). In an organism, those cells other than germ cells.

Spontaneous Mutations. Mutations whose cause is not "observable."

Transcription. RNA synthesis by base pair matching with DNA.

Transduction. Process by which bacterial genes are transferred by a bacteriophage particle from one bacterium to another.

Transformation. Purification of DNA from cells or viruses and incorporation of it into a cell in order to induce genetic modification there.

Translation. Direction by mRNA of the sequencing of certain amino acids; occurs at protein synthesis.

Translocation. Process in which part of a chromosome is transferred into another part of either a homologous or nonhomologous chromosome.

Triploid. Having three times the haploid number of chromosomes found in the nucleus; triploidy is a form of polyploidy.

Turgor. State of a cell when there is a rigid cell wall stretched by an increased volume of vacuole and protoplasm at absorption of water; cell is then turgid.

Turnover Number. The quantity of substrate molecules transformed per minute by an enzyme working at its maximum rate.

Zygote. Fertilized ovum prior to cleavage.

Index